U0054300

西班牙人常説，
愛情是任何一個食譜的基本材料…

In Spain we use to say :
"love is the basic ingredient of any recipe…"

西班牙有許多不同的省份地區，各有不同的料理手法和代表菜。

在這本食譜書裡，我嘗試把每個地區的一小部分都納入進來。

我想傳達的東西很簡單——料理其實是一種樂趣！

In Spain we have many different regions, each one has her own way of cooking, that's why I try to introduce you a small part of each region.

What I would like to introduce you on this book is something really simple, cooking is a pleasure!

目錄
Contents

靈感乍現的創意菜
My Own Creations

融合的美味
Fusion

以美食驚艷西班牙

我是西班牙觀光局駐東京辦事處主任，台灣、日本與南韓三地的旅遊市場都屬於我的辦事處管轄範圍內，因此我很高興能替丹尼爾在台出版的第一本食譜為文作序 。

對西班牙人而言，台灣依然是個遙遠、陌生的所在。同樣地，對台灣人而言，西班牙也是一個遙遠、陌生的地方。不過，傳統料理與新式美食在這兩個國家之間搭起一道文化的溝通橋樑。主廚丹尼爾雖然年輕，在推廣新式美食之餘，卻不忘向傳統的西班牙料理致敬，在同行之中備受推崇。這本食譜反應出他在烹飪上的經歷（時間雖短卻愉快）、對美食的理念與精巧古怪的食譜，也代表這派烹調所提出來的新焦點。

位於地中海上的西班牙是個得天獨厚的國家，一年四季氣候溫和，生產製造出各種不同的產品，如初榨橄欖油與美酒皆聞名全球。西班牙何其幸運，擁有豐富的釀酒文化與美食文化，複雜多變一如其境內變化多端的分區；高檔美食有走前衛的，也有地方風味餐，如瓦倫西亞的海鮮燉飯、安達魯西亞的番茄冷湯、加泰隆尼亞的海鮮麵，如今都被視為西班牙的地中海餐，聞名於世。

丹尼爾花了兩年的時間在台灣當主廚，和大家分享他的料理創作，努力提高西班牙這個國家的形象，因為這個國家雖然擁有非凡的歷史，卻不為台灣人所熟知。所以，我們要感謝這位充滿創意的主廚所做的努力，他透過烹飪藝術和美食推廣西班牙文化，這種美食逐漸受到新一代食客的歡迎。

雖然我身在東京，但是身為西班牙高檔美食的愛好者，我要預祝丹尼爾成功，這是理所當然的，同時也要感謝他的努力與樂觀。本書的讀者，希望你們能夠從這位大師的新式食譜之中得到樂趣，嚐嚐新式的融合菜，享受新的體驗和主廚高明的烹飪技巧。至於有雄心壯志的廚師，則不妨把這本食譜當做家鄉菜不可或缺的手冊。

我更鼓勵大家赴西班牙旅遊，享受西班牙歷史悠久的佳餚或新興的美食，相信不論哪一種都會令你驚艷。

西班牙觀光局駐東京辦事處主任

杜卡斯

copyright from ICEX

Experience Spain with gourmet

As director of the Spanish Tourism Office in Tokyo (Japan), it is a pleasure for me to be selected to write the preface of Daniel Negreira's first cook book published in Taiwan—which, along with Japan and South Korea—is one of the three tourist markets managed by my office.

For Spanish people, Taiwan is still a distant, little-known destination. Similarly, the same is true of Spain for the Taiwanese. However, traditional cuisines and the new gastronomy are working as an important cultural bridge between the two countries. The master chef Daniel Negreira, in spite of his youth, holds an esteemed position among the select chefs working to promote a new gastronomy while still honouring traditional Spanish cuisine. This book represents the new focus of that school of cooking, as reflected in brief and pleasant histories about Negreira's culinary experiences, gastronomic ideas, and curious recipes.

Spain is a privileged Mediterranean country with mild temperature almost year round. This allows various world-renowned products to be cultivated such as virgin olive oil and excellent wines. Spain is fortunate to possess a rich oenological and gastronomic culture, as diverse and complex as its varied regions, with an established high cuisine featuring both avant-garde and popular dishes like the paella valenciana, gazpacho andaluz, and fideua catalane—now associated with the Spanish Mediterranean diet known throughout the world.

Daniel Negreira spent two years practising his profession in Taiwan, sharing his new creations with all in an effort to enrich the image of a country that, despite its remarkable history, is still relatively unfamiliar to many Taiwanese. Therefore, we should be grateful for the effort of this creative chef to promote the culture of Spain through his culinary artistry and a cuisine that is increasingly popular with a whole new generation of diners.

From Tokyo and as Spanish lover of high gastronomy, I wish Daniel all the success that he deserves and thank him for his efforts and optimism. For you, dear reader, I hope you enjoy the new recipes of this grand master. Taste the new fusions, enjoy the new experiences, and delight in his excellent cuisine. I invite aspiring cooks to consider this book to be an indispensable manual of home recipes. And, certainly, I encourage you to travel to Spain and to keep enjoying our marvellous, millenarian and young gastronomy.

Director, Japan, Korea & Taiwan National Tourist Office of Spain
Ignacio Ducasse

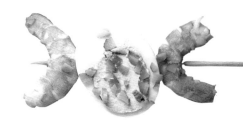

西班牙大廚在台灣

偉大的西班牙作家坎桂羅（Alvaro Cunqueiro）曾經說過：「雖然大家都會吃，但是沒幾個人懂得怎麼吃。」

我們應該用心選擇要吃的菜，用想像力品嚐這些菜，如果沒有想像力，享受一頓美食的樂趣不如簡化成吞服幾粒藥丸。坎桂羅那句話的意思是，保持心靈開放的人才能真正享受食物的滋味。這不表示做菜一定要用高價的食材，因為即使是便宜的東西也能變成美味的食物。

飲食只是為了填飽肚子，我反對這種想法；我認為料理表現的是精神，它是探索自然的祕密所呈現出來的極致藝術。希臘人認為廚師是追求和諧的藝術家，這點和中國人所謂的均衡觀念不約而同。一個廚師必須懂得食材，知道如何混用不同的食材，搭配出均衡的食譜。廚師還要夠大膽，避免一再重複同樣的手法，他必須尋找方法讓消費者感到驚奇。廚師必須和他所使用的食材相合。

美食一直是西班牙地中海沿岸文化重要的組成，也成了我們身分認同的一部分。雖然地中海沿岸不同地區的人差異頗大，但是我們的食物，都是由麵包、美酒和橄欖油這三種神聖的食材所組成的。這三種食材可以搭配山產或海味，做出既美味又健康，人人都適吃的菜。地中海飲食不僅有益健康又能延長壽命，非常有趣。眾所皆知，葡萄牙人在十七世紀占據澳門，過起舒舒服服的日子，他們看見台灣，稱這座美麗的島為福爾摩莎（Formosa），不過葡萄牙人從未登上這座寶島。倒是西班牙水手登上了這座島，稱它為福爾摩莎島（Isla Hermosa）。西班牙人在此殖民定居，建立聖地牙哥堡和聖多明尼各堡。

沒有證據顯示，在這群西班牙水手之中有廚子；事實上，一直要到幾個世紀後，主廚丹尼爾來到台北，台灣人才算等到了。雖然早在丹尼爾之前就有西班牙主廚來到台北，但是沒有一個人的成就及得上他。原因有可能出在之前來的主廚都把重心放在「海鮮飯」和「小點」等傳統料理上，於是乎他們開的餐館逐漸為人所淡忘，他們所提供的菜色也跟出現在此地的其他國籍料理混在一起。

丹尼爾則不一樣。他在這裡端出來的，不是我們在西班牙吃了幾百年的傳統菜。雖然他是以傳統為基礎去創作新菜，但是他所感興趣的不僅是傳統料理，他從古老的食譜汲取知識，再去轉換、改良與創新，結果端出來的是美

Spanish chef in Taiwan

The great Spanish writer Alvaro Cunqueiro once said, even though all the people eat, just a few of them know how to eat.

We should use our mind to choose which dishes we will taste, and taste it using our imagination, because without imagination, the pleasure of a nice meal might as well be reduced to a few pills. What Cunqueiro means is that those who really enjoy food are those who are open-minded. This doesn't mean that we must use costly ingredients, since even the cheapest products can become something delicious.

Against the belief that food is just to fill the stomach, I believe cuisine is something connected with the spirit and a supreme art that seeks to probe the secrets of nature. Even the Greeks considered chefs to be artists working towards harmony, something that matches the Chinese concept of balance. A chef must know the ingredients, how to mix them, and how to craft a balanced recipe. But he also needs to be bold enough to avoid repeating the same approach; he must look for ways to surprise the customer. The chef must be in harmony with the ingredients he is using.

Gastronomy has always been an essential part of culture in our region of the Mediterranean, and it has become part of our identity. Even though there's a big difference between people in the various regions around this sea, we share a miraculous combination: the holy trilogy of bread, wine and olive oil. This trilogy can be combined with products coming from the sea or mountains to obtain something delicious and healthy for everyone. The Mediterranean diet is fun, healthy, and life-prolonging. It's well-known that during the 17th century, Portuguese people gazed at Taiwan from their comfortable holdings in Macao and described it as FORMOSA, but they never landed on this fantastic island. However, the Spanish mariners did, calling the isle Isla Hermosa. They established settlements, building Fort Santo Domingo and Fort Santiago.

There is no evidence that among those mariners there was any chef; in fact, the Taiwanese people needed to wait a few centuries until the young chef Daniel Negreira arrived in Taipei. Even though other Spanish chefs arrived in Taipei prior to Daniel, none met with the success that Daniel did. Maybe this is because they only focused on the basic, traditional recipes such as paella and tapas, consigning their restaurants to oblivion and mixing their menus with other national cuisines already found in Taiwan.

Daniel's case is different. He didn't came to offer us the traditional recipes we ate in Spain for centuries. He's not only interested in that, even though he uses tradition as a base to create something new. Using

味的食物。他之所以能夠做到這一點，不只是因為他天賦異稟，更因為過去在西班牙他跟在一流的主廚下面做事，且曾經周遊各地：他曾經共事過的對象有米其林三星名廚 Martin Berasategui、EL BULLI（2006、2007、2008 連續三年被選為全球最佳餐廳）的老闆兼主廚阿德里亞（Ferran Adria) 和三星餐廳 Akelarre（全球排名第八的餐廳）、三星餐廳 Mugaritz（2007 年全球排名第七、2008 年全球排名第四的餐廳）的主廚等，最後還當上拉托哈島 (La Toja) 上的 Gran Hotel Hesperia 五星級飯店主廚。

西班牙美食和法國或義大利美食有一點不同，那就是一旦離開西班牙，就很難備妥食材。這個問題出在西班牙菜的重心不僅在於它的烹調方法，材料更是它的關鍵。西班牙的肉類、海鮮和蔬菜都很特別，在台灣不容易找到。

不過，丹尼爾可以設法去面對這項挑戰。他對本地市場做過一番徹底的研究與搜查，不僅找到需要的食材，甚至採用本地的食材做出西班牙菜，而不失去其原味。有時候，他還會改良食譜，成果就是神奇的融合菜。丹尼爾的這些料理不過是個開端而已，它將會經年累月持續下去，成為「丹尼爾的料理」。

近來，作菜已經從家庭主婦為了餵飽一家人所做的無趣事，轉變成人人都能夠感興趣的事。漸漸地，烹飪成了一種嗜好，一種放鬆的方法，變成消遣或以食會友的一種活動。

你在這本書裡看到的食譜不論是在家烹調，或是在大膽挑戰美食的聚會場合都很適合。這些食譜雖然採用西班牙菜的烹調技巧，但是每一道菜的設計都很簡單，不論什麼人，對食物和烹飪器材的認識再怎麼少，都做得來。這些菜吃起來跟你到任何一家高級餐廳吃到的一樣令人滿意，只不過是在家裡做出來的而已。

讀者可以根據這本書的設計，選擇採用傳統料理的方式或新派改良版的方式去烹調。你可以體驗不同的烹調方式，也可以比較食譜的差異，同時精進你的烹調技巧。丹尼爾想用這本書介紹我們認識最新的美食創意，包括結合亞洲食材、西式技巧與傳統食譜的融合菜。

談夠了。那麼，該進入美味的西班牙料理世界了。走進廚房，開始忙吧！

台北西班牙商務辦事處處長
拉瑪斯

these age-old recipes as a source of knowledge, he shifts, upgrades, and innovates. The final result is something delicious. He is able to do this not only because his natural gift, but also because of his training with the best chefs in Spain and in other countries around the world. For example, he has experience with: Arzak 3* Michelin--considered the 8th best restaurant in the world; Martin Berasategui 3* Michelin; Ferran Adria, owner and chef of EL BULLI 3* Michelin--classified as the best restaurant in the world in 2002, 2006, 2007 and 2008; Akelarre 3* Michelin; Mugaritz 3* Michelin--considered the 7th best restaurant in the world in 2007 and 4th in 2008; and, finally, he has experience as chef in the Gran Hotel Hesperia 5* GL in La Toja island.

Spanish gastronomy, in contrast to French or Italian, is something that is not that easy to prepare once you are not in Spain. This is due to the fact that Spainish food is focused on the ingredients, not only on the way we cook it. The products from Spain such as the meats, seafood, and vegetables are so unique that they are not easy found in Taiwan.

But for Daniel, it is a challenge that he can manage. After doing exhaustive research on the local markets, not only was he able to find the products, he even adapted some local products to fit the recipes from Spain without losing any quality. Sometimes he even upgraded recipes. The final result is a fantastic fusion--something that is only beginning, but will last for years: Daniel´s cuisine.

Recently, cooking has shifted from being something done by a bored housewife to feed the family to something that everyone can take an interest in. Increasingly, cooking is a hobby, a way to relax, or an activity that can amuse and bring people together around food.

The recipes you will find here will serve well in the home or at the boldest culinary gatherings. Using Spanish culinary techniques, the recipes are all designed to be easy enough that anyone can prepare them with a minimum knowledge of food and home appliances. The dishes will be as pleasing as any that can found at a high class restaurant, but prepared at home.

The way the book is designed, the reader can choose to follow the traditional or the modern version. As a result, you can sample different cooking methods, as well as compare recipes, while sharpening your skills in the culinary arts. With this book, Daniel is trying to introduce us to his latest culinary creations, including some fusion recipes that combine Asian products with western techniques and traditional recipes. Enough talking. Now it´s time to enter the delicious world of Spanish food. Let´s go into the kitchen and get busy!

Director of Spanish
Chamber of Commerce
José Luís Lamas

推薦序　Foreword
主廚另一半的幸福食驗

　　身為丹尼爾的另一半，實在很難提筆替他的第一本食譜作序，因為他是個出色的主廚，我根本不知道該如何替他的食譜增色。 因此，與其嘗試在他的「專業」領域方面替他畫蛇添足，我還不如來談談他的私生活。我們認識沒多久以後就決定在一起。老實說，看到他的第一眼我就被他吸引了，原因不只是他的外表，同時也因為做為一個主廚與做為一個男人的他所表現出來的那份優雅與自信。事實上，他改變了我對傳統廚師的印象，過去我一直認為廚師就是身上圍著一條油膩膩的圍裙，長得肥滋滋的老傢伙。

　　他跟我聊起他的專業與工作經驗，知識之淵博令我印象深刻。同時，我也很開心。我告訴自己：「我真幸運，能夠交到一個主廚男友，隨時可以做飯給我吃！」時光飛逝，我心中的想法漸漸成真。隨著我們之間的關係進展，我不僅從他那裡學到西式的烹飪技巧，也逐漸明白烹飪其實一點都不難。相反的，烹飪十分有趣！

　　想當年，我們還住在西班牙，想要吃中國菜的時候，我就會去買材料來做。做中國菜準備起來既複雜又費時，這點大家都知道，如果是只做給一、兩個人吃，更是不經濟。但是，這位神奇的大廚教我利用冰箱裡現有的食材，迅速做出美味的菜餚，他是我的救星！舉個簡單的例子，用罐頭鮪魚、玉米粒、洋蔥絲、橄欖油和芝麻醬油拌什蔬，就可做出一道超級美味的中式沙拉！除此之外，這道融合菜讓我覺得我也可以成為一個有創意的主廚！

　　從我們相識的那一刻開始到現在，丹尼爾一直都在做菜給我吃，有我熟悉的食物，也有手法令人稱奇的菜色。例如，有一次我病了，早上丹尼爾把我叫醒，在我身旁擺了一頓豐盛的早餐。他就是這個樣子！對我而言，他在工作表現上既專業且充滿了創意，私底下則是一個聰明又體貼的丈夫。

　　現在，不僅是我可以享受到丹尼爾做的菜，我們的寶貝女兒瑪莉娜也可以分享他的創造力。我們的女兒雖然才一歲大，卻已經嚐過她爸爸做的鴨肝與松露等精緻複雜、美味可口的菜，而不是一般的嬰幼兒食品！也許是因為吃西班牙菜的關係吧，她長得健健康康、活潑好動且聰明伶俐。丹尼爾的創意菜裡絕對是飽藏著一個父親對女兒的愛。我和瑪莉娜真是何其有幸！這個了不起的主廚將不斷地為給我們帶來樂趣，也是他的妻子和女兒學習的對象。

莫妮卡

The Fantastic Experience of Being Wife of Chef Daniel

Being Daniel's wife, it's hard to write the preface for his first book, as I have no idea how to contribute to this outstanding chef's recipes. Therefore, rather than attempting to add anything professional, I will talk about his personal life.

We decided to be together very soon after we met. And, speaking sincerely, he did attract me at first sight, not only because of his appearance, but also for the elegance and the self-confidence that he expresses as a chef and as a person. In fact, he did make me change my impression of the traditional chefs, as I used to think of them as being fat, old guys in greasy aprons.

When he talked with me about his profession and his working experiences, I was impressed by his knowledge; but, at the same time, I was so happy. I told myself, Hmmm how lucky I am that I can have a boyfriend who is a chef that can cook for me always! Little by little, as time goes by, what I had in mind has been being realized. As our relationship developed, I also learned from him western cooking skills and have come to realize that cooking is not hard at all. On the other hand, it is such an interesting skill! When we still lived in Spain, whenever I wanted to have Chinese food, I used to go buy the ingredients for cooking. As everyone knows, preparing Chinese food can be quite complicated, time consuming, and—if cooking for only one or two people—not very economical. However, my saviour, the magic chef, taught me to utilize whatever was left in the refrigerator to create a fast but delicious dish. One simple example is vegetables with canned tuna fish, corn grains, sliced onion, olive oil, and soy with sesame sauce. The result is a super delicious salad in Chinese style! In addition, it is a fusion dish that makes me feel that I can also be a creative chef, too!!!

From the moment we met until now, Daniel has been cooking for me, both familiar foods and, sometimes, some recipes with a surprise. For example, once when I was sick, Daniel woke me in the morning with an abundant breakfast by my side. This is him! For me, he is professional and creative at work, smart and thoughtful as a person and a husband.

Now, not only can I enjoy Daniel's cuisine, but our lovely daughter Marina can also share in his genius. Only one year old, she has already tasted her daddy's delicious and sophisticated dishes, like duck liver and truffles—not your normal baby food! She is so healthy, active and smart, perhaps because of the magic of Spanish cuisine. Daniels's culinary creations certainly contain a father's love for his daughter. How lucky Marina and I are! We won't stop enjoying and learning from this grand chef: dad and husband forever.

Monica

一起 Fun 做菜

在本書中，我想傳達的東西很簡單——料理其實是一種樂趣！當我們偶爾出外用餐時，看起來很複雜的料理實際上是很容易準備的。通常我們需要的只是一點點耐心、一些基本知識和基本工具。但最重要的是要有熱情，這是在烹飪中最重要的元素。事實上，在西班牙有句俗諺是這樣說的：「愛情是任何食譜的基本材料…」。

你將在這本書中發現好幾種不同型態的食譜：傳統 vs. 創新、我的創意菜，和我從台灣本地廚師身上學習到又再經轉化的「融合的美味」。西班牙有許多不同的省份地區，各有不同的料理手法和代表菜。在這本食譜書裡，我嘗試把每個地區的一小部分都納入進來，例如東半部的西班牙海鮮飯、西部的章魚、北方的雞肉和南邊的冷湯等等。

現在讓我們來趟美食之旅吧，藉此我希望烹飪是有趣的，而且讓你感覺料理美食真的——並非難事！

What I would like to introduce you on this book is something really simple, cooking is a pleasure, and sometimes we go to a restaurant and order some dishes that seem to be really complicated, but in fact it's really simple to prepare.All we need is a little bit of patience, some basic knowledge and a basic equipment.The most important thing is to have some passion and include this on our recipes as the main ingredient, in fact in Spain we are use to say : "love is the basic ingredient of any recipe…"

Through this book you will find different recipes, traditional way combined with the modern version, my own creations and some fusion recipes that I try to prepare learning from the local chefs about the Taiwanese food and changing it into something completely new. In Spain we have many different regions, each one has her own way of cooking, that's why I try to introduce you a small part of each region, from the east style paella to the west style octopus, the north style chicken or the south style cold soup.

My intention is to drive you through a gastronomic journey, trying to make it fun and showing you that cooking is really easy.

主廚 丹尼爾·尼格雷亞
Chef Daniel Negreira

「EL　TORO」不同於其他餐廳的地方，是主廚丹尼爾非常於勇於創新，不斷尋找新食材和新做法，不做一成不變的菜，因此能充分滿足老饕客人的需求。除了有一身好手藝，丹尼爾的體貼和用心也讓我印象深刻，今年過年前夕我向餐廳訂了豬腳外帶，由於服務生的疏忽，豬腳未烤到理想熟度，讓我大為光火，丹尼爾聽到吵鬧聲跑出來，問清原由後，立刻道歉並承諾會把烤好的豬腳親自送到我家。對丹尼爾來說，我只是一個發牢騷的客人，但他處理危機的態度讓我留下深刻印象，也讓我們不「打」不相識，從此變成他的常客。

德意志銀行　Eric Lin

The difference about EL TORO from other restaurants is that Chef Daniel is not afraid of innovating. The continuous searching for new ingredients and cooking, and no invariable cuisine allowed have completely satisfied the demands of the gluttons.

Apart from the excellent cooking, Daniel's consideration and attention also make me deeply impressed. This New Year's Eve I ordered a take-out of the pork knuckle at the restaurant. Because of the negligence of the waiter, the knuckle was not grilled to the degree I desired. I was annoyed and Daniel heard the fighting and ran out to ask the reason. He apologized right away and promised that the finished grilled knuckle would be delivered personally to my home. For Daniel, I was just a grumbling customer. But the attitude he dealt with the crisis left me a great impression. Fighting has made us be acquainted with each other and hereafter I became a frequent visitor of his.

Eric Lin, Deutsche　Bank

▲

一年多以前，透過愛吃的朋友介紹，認識了丹尼爾的手藝，這位年輕的西班牙大廚端出來的菜餚讓我相當驚艷。後來經常約朋友到「EL　TORO」打牙祭，幾次之後我連菜單都不用看，直接請主廚幫我們搭配菜色，丹尼爾很用心，每次配出來的菜色都不同，而且總是能讓賓客盡歡。

亨信公司負責人　張錕章

I have known Daniel's cooking through some friends, who have a special fond in food since a year or more ago. The dishes this young Spanish chef serves have always quiet astonishing. Therefore, we always go to EL TORO to have some feasts with friends. After a number of times, I don't even have to look at the menu and we will have the chef design the menu directly. Daniel is very attentive; every time the menu designed are all different and always make our guests full of complete satisfaction.

Kun-Chang Chang, Head of Trustwell Co. Ltd.

「EL　TORO」擁有一間神奇廚房，端出來的菜色總是那麼特別，主廚把不同的食材和味道結合在一起，透過嶄新的擺盤，讓客人有耳目一新的感覺，在台灣很少見到這麼有趣的嘗試。

Hung Chains International Co.,Ltd
Paul Wyss

EL TORO owns a marvelous kitchen. The dishes served are always quiet unique. The chef has combined different ingredients and savors through the brand-new dish designs to make the customers to feel new and fresh. You don't quiet see such an interesting challenge here in Taiwan.

Bailli Délégué de Taiwan La Chaîne Des Rôtisseur
Paul Wyss, Hung Chains International Co.,Ltd

丹尼爾是我的女婿，我視他為半子，他的手藝好無庸置疑，但我更肯定的是他的認真和用心。餐廳剛開幕那一陣子，他凡事親力親為，從無到有，整整一個多月，沒日沒夜打拚，這種認真做事的態度讓我相當感動，也是「EL TORO」成功的關鍵。

丹尼爾的岳父 & EL TORO 的老闆
魏漢隆

Daniel is my son-in-law, whom I consider as my half-son. His cooking is, without a doubt, one of the best in the culinary world. Although what I recognize from him the most is his total commitment and concentration. At the beginning of the restaurant opening days, he took part in organizing everything from scratch. For one whole month, he worked day and night. Such attitude of full commitment has highly touched me and it is also the key to success for " EL TORO ."

Han-Long Wei,Daniel's Father-in-Law&Boss of EL TORO

身為葡萄酒講師，我經常為客人舉辦餐酒品嚐會，辦了這麼多年餐會之後，只有少數餐廳是我在去之前還會興奮與期待的，「EL　TORO」就是其中一家。我喜歡這裡的菜，因為它不落俗套，主廚丹尼爾總是能為每一次光臨的客人帶來許多飲食上的驚喜。

葡萄酒講師 劉鉅堂

As a lecturer of the wine, I usually hold the wine tasting gatherings for the customers. After so many gatherings over the years, the restaurants that I will feel excited and look forward to are quiet a few and EL TORO is one of them. I love the food here because it's not conventional. Chef Daniel can always bring surprises on the food for the every time visiting customers.

Jason Lau , Wine lecturer

丹尼爾的菜很棒，因為他對食材的品質要求非常高，大自肉類、海鮮、蔬菜，小至橄欖油、香料、鹽巴，無一不是精品，加上他的盤飾賞心悅目，讓用餐變成一種享受。

西班牙大使夫人 Casilda Baeza Gomez

Daniel's dishes are great, because he has quiet a high request on the quality of ingredients. As big as from meats, seafood, vegetables and to as little as olive oil, spices and salt. Everything must be exquisite. Moreover, his garnishes are pleasing and make dining to become an enjoyment.

Casilda Baeza Gomez,
Wife of the Director of Spanish Chamber of Commerce

第一次吃丹尼爾的菜，是在台北西班牙商務辦事處處長舉辦的餐宴上，他邀請這位資歷豐富的西班牙新生代主廚為大家掌杓，讓我留下深刻的印象。一個星期後，我跑到丹尼爾新開的餐廳用餐，從此變成這裡的熟客。

丹尼爾很重視食材的新鮮度和品質，同時他對食物和烹飪充滿熱情，這是最感動我的地方，吃他的菜彷彿遊歷西班牙各個省份。我曾在歐洲長住 20 年，對西班牙菜頗多涉獵，吃過許多地方的西班牙美食，我敢說「EL TORO」的西班牙菜是全亞洲最道地的選擇。

Barclays 銀行總經理 陳冠達

When I first had Daniel's dish, it was in a feast held by the director of Spanish Chamber of Commerce, Taipei. He invited this experienced Spanish chef of the New Generation to be in charge. I was left with a great impression. I went to have a meal at Daniel's new opening restaurant a week after and have become a frequent visitor here.

Daniel places a great emphasis on the freshness and quality of the ingredients. He's also full of passion for the food and cooking at the same time. This is what touches me the most. Having his dishes is like making a journey of all the provinces in Spain. I have stayed in Europe for 20 years and been quite familiar with the Spanish cuisine. Having experienced with many Spanish cuisine from all over the places, I will say EL TORO is one of the most authentic choices for Spanish cuisine in all Asia.

Kuan-Da Chen, Manager of Barclays Bank

我在上班途中意外發現「EL　TORO」的時候，這間餐廳才剛開沒有多久，週末我帶著太太小孩上餐廳「探險」，非常意外吃到了正宗伊比利火腿。

　　後來我變成餐廳的常客，發現這裡的菜不像一般台灣常見的西班牙餐廳，清一色賣著海鮮飯和炸丁香魚一類的 Tapas 小點，反而比較像我在西班牙拜訪過的現代餐廳。主廚丹尼爾經常創作發表新菜，有時候還能吃到當紅的分子廚藝概念菜，我把餐廳推荐給朋友，吃過的人也都讚不絕口。

TVBS 週刊發行人　邱一新

　　When I accidentally ran into EL TORO on the way to work, this restaurant has just begun its business. I took my wife and children to the restaurant for the exploration over the weekend and very surprisingly we had our genuine Iberian ham.

　　I have become a frequent visitor of the restaurant later on. I found that the dishes here are not commonly seen in the Spanish restaurants of Taiwan. Instead of selling variety of appetizers, tapas, like paella and fried lilac fish, it's more like the modern restaurant I've visited in Spain. Chef Daniel frequently creates and publishes new dishes. Sometimes, you can even have the hottest concept of molecular gastronomy cuisine. I recommended the restaurant to the friends and none of them are disappointed.

Yi-Hsin Chiu, Publisher of TVBS Weekly

　　我是主廚丹尼爾的粉絲，我認為「EL　TORO」是目前我在台灣吃過最棒的西班牙餐廳，這裡不但能吃到最多樣的 Tapas，最讓人興奮的是還可以直接跟主廚討論菜色，關於烹飪的靈感好像藏在他的腦袋裡源源不絕，這讓「EL TORO」變成一個挖之不盡的美食寶藏，絕不僅限於小小一張菜單。

Managing Director Trumpf Taiwan IndustriesCo.,Ltd
Wolfgang Vranze

　　I am a fan of Chef Daniel. I think EL TORO is the greatest Spanish restaurant I've ever had so far in Taiwan. You will not only have the most diversified tapas here, what excites me the most is that you can directly discuss the dishes with the chef. The inspiration about cooking is like hiding in his brain and never ends. This has made EL TORO to become a gourmet treasury unable to be dug out. A little piece of menu won't tell you all.

Wolfgang Vranze, Managing Director Trumpf Taiwan Industries Co.,Ltd

傳統 vs 創新
Traditional Food & Modern Food

相同食材，不同呈現，一個是經典，一個是創新，
都是多樣、繽紛又健康的西班牙菜。

A compilation of variety, color and healthiness,

the base of our diet…with two different points of view,

using the same ingredients, a classic way and a new Spanish food version.

Leeks Soup with Smoked Clams
Soft Gelatin

蒜苗湯佐煙燻蛤蜊凍

西班牙北方冬季嚴寒，氣溫經常急降到攝氏零度以下，入冬之後，類似中國火鍋或砂鍋的Hot Pot相當風行，人們靠熱騰騰的Hot Pot驅走寒意。

「蒜苗蛤蜊鍋」是一道很經典的Hot Pot，最美味的品嚐時間，是二月青蒜當令的時候，西班牙冬春之交盛產的青蒜，清甜多汁，味道美極了。我到台灣之後，發現這裡的青蒜沒有西班牙那麼幼嫩甜美，但只要加一點糖，也可以煮出一樣的好味道。此外，還可以用南瓜取代胡蘿蔔來提升甜味。

一樣的砂鍋材料，我試著更換另一種呈現方式，利用蛤蜊高湯做成蛤蜊凍，淋上濃稠熱燙的蒜苗湯，讓蛤蜊的鮮甜在熱湯催化下，慢慢融化、釋放⋯，很傳統的老湯鍋馬上變成充滿摩登食趣的「蒜苗湯佐煙燻蛤蜊凍」。

It's bitter cold in Spain in winter; the temperature often drops to below zero degree Celsius. When winter comes, hot pot is quite popular. People eat hot pot to disperse cold.

Clam casserole with garlic sprout is a classic hot pot. The best time to taste it is in February when green garlic is in its season. Green garlic harvest in late winter or early spring is so fresh, so sweet and succulent that it tastes marvelous. Being here in Taiwan , I found that local green garlic is not as fresh and sweet as that in Spain. However, with just a little bit of sugar the flavor is good as well. Besides, you can substitute pumpkin for carrot to enhance its sweetness.

With the same ingredients, I try to present it in a different way. First, I make clam jelly with clam stock and then pour the rich green garlic soup over the clam jelly. The jelly will slowly melt and release its sweet aroma under heat. Immediately, a traditional casserole has become a modern dish full of fun.

Cooking time ›› 20 minutes.

主材料：

蒜苗	1公斤	1 kilogram leeks
蛤蜊	500公克	500 grams clams
馬鈴薯	2顆	2 potatoes
南瓜	150公克	150 grams pumpkin
荷蘭芹	20公克	20 grams parsley
0′4° ANTONIO CANO 橄欖油	25公克	25 grams olive oil (0′4° ANTONIO CANO brand)
無鹽奶油	50克	50 grams butter (non-salted)

新派做法材料： / **Modern**

明膠片（吉利丁）	32公克（約4片）	32 grams gelatin (4 leaves)
鹽	5公克	5 grams salt

傳統 Traditional 蒜苗濃湯做法：

1. 將蒜苗切成4公分長段，只取中間較嫩處，其他部分摘除。只用蔥白和一點點蔥綠部分。

2. 馬鈴薯削皮切成 4 x 4 公分小塊；南瓜削皮切丁成同樣大小。用一只大平底鍋加熱奶油和橄欖油，放入南瓜、馬鈴薯和蒜苗，煮約2分鐘，加入清水蓋住全部材料。

3. 煮滾後用慢火煮至馬鈴薯變軟，加入蛤蜊，再用大火煮滾至蛤蜊殼打開。蛤蜊煮開後，汁液本身就有鹹味，嚐過後若覺得不夠鹹，可再加鹽調味。

4. 盛盤後撒上荷蘭芹點綴。

Cut the leeks into pieces 4 cm. long, using only the inside leaves. Trim so you are using mainly the white stalks tipped with a bit of the green.

Peel the potatoes and dice into 4 cm. x 4 cm. cubes. Peel and cube the pumpkin to the same dimensions. Heat the olive oil and the butter in a large pan, then add the pumpkin, potatoes and the leeks. Cook for two minutes and then cover it with water. Bring to a boil and cook until the potatoes are tender, then add the clams and boil all together until the shells are open. If needed, add some salt after the clams are open, since the water inside the clams is usually salty enough.

Garnish with chopped parsley.

新派 Modern 蒜苗湯佐
煙燻蛤蜊凍做法：

1. 將蒜苗、馬鈴薯和南瓜切成如同傳統做法中一樣大小，但只先煮蒜苗。煮約2分鐘後放入馬鈴薯，再加水蓋過所有材料，煮至馬鈴薯變軟。

2. 取出所有材料用果汁機打勻過濾。加入約5公克鹽調味。

3. 在一只大鍋中加水蓋過切好的南瓜，慢煮至南瓜變軟。將煮熟的南瓜和一半煮南瓜的水，一起放進果汁機中打成乳白色。將明膠置於容器中，放入冷水和冰塊讓明膠軟化。

4. 將2片明膠放入南瓜汁用打鮮奶油器（一種專門用來打鮮奶油的機器）打勻。利用打鮮奶油器與空氣壓縮將N2O（一氧化二氮）打入南瓜汁中充氣之後，放入冰箱冷藏2小時。

5. 在另一只鍋中，用20cc的水將蛤蜊殼煮開（煮時鍋子用鍋蓋蓋住，以免水分蒸發）。之後用牙籤，每3個蛤蜊肉串成一串。

6. 將煮蛤蜊的高湯過濾之後，倒入冰涼的明膠中混合，之後再倒在蛤蜊上將蛤蜊蓋住，放入冰箱冷藏2小時。

7. 將蛤蜊凍切成圓形，拿掉牙籤之後放在盤子底部，用1片荷蘭芹放在上面做裝飾，之後挖一匙南瓜泡放在蛤蜊凍旁邊。上菜的時候，將滾燙的韭蔥馬鈴薯湯淋在蛤蜊凍上，輕輕攪拌，讓它慢慢溶化⋯⋯

Cut the leeks, potatoes, and pumpkin the same way as you did for the traditional recipe, but begin with just leeks in the pan. After cooking for two minutes, add the potatoes and cover with water. Let it cook until the potatoes are tender. Next, blend the mixture and strain it. In this version, add some salt. Five grams should be enough.

Cover the pieces of pumpkin with water and cook slowly in a large pan until tender. Afterwards, retain half of the cooking water and blend together with the pumpkin to obtain a light cream.

Get the gelatin leaves and put in a bowl with cold water and some ice cubes until the mixture is soft.

Add two leaves of the soft gelatin to the pumpkin cream and whip with a cream whipper—a tool specifically designed for whipping cream. Using the compressed gas feature (N2O) with the cream whipper, aerate the cream and put inside the refrigerator for two hours.

In another pan, steam open the clams using 20 grams of water. Keep them covered at all times in order to minimize evaporation. Skewer the clams with a toothpick, three clams per stick.
Strain the liquid left over from steaming the clams. To the chilled gelatine, add the warm clam juice. Cover the clams with the clam gelatine and put inside the refrigerator for two hours.

To serve, cut the pieces of clam gelatine into preferred shapes. Remove the toothpick and place on the bottom of the dish. Garnish with a leaf of parsley. Spoon the pumpkin foam beside the gelatine and then, with guests seated at the table, serve the hot leek and potato soup over the clam gelatine in order to melt it. Mix and ....yummmmmm!!!!

Tips：
如果把海鹽換成法國產的煙燻鹽，那麼蛤蜊凍裡就會帶有淡淡的煙燻香氣，提供嗅覺更多一重玩味。
To add an extra smell, substitute French smoked salt for sea salt, and you can smell a whiff of smoked flavor from the clam jelly.

Seafood Soup

海鮮湯

　　西班牙海岸線長，許多小漁港邊都有好喝的海鮮湯，魚骨蝦殼煉出來的鮮美湯頭，滋味真是美極了！在此貢獻兩種海鮮湯食譜，一種遵照傳統，是做法很簡樸的海鮮湯，僅利用香料和蔬菜吊出海產鮮甜。

　　另一種則玩弄少許料理技巧，利用蛤蜊高湯以鮮奶油發泡機打出泡沫來，再把分門別類處理的海鮮，如煎得酥脆的淡菜、Q彈的蝦球裝盤，注入海鮮高湯，再放上一朵用蛤蜊高湯打出來的奶泡，享受味蕾上多層次的美妙變化。

A lengthy coast in Spain has enabled a lovely seafood soup around the many fishing ports, with fresh and tasty soup stewed from fish bone and shrimp shell, a lovely taste than ever! I hereby contribute two recipes for seafood soup, one follows tradition, a seafood soup with simple and easy manner of cooking, which only uses spices and vegetables to enable the sweetness and freshness of seafood.

While the other is to play with little culinary skill; using clam soup stock with bubbles whipped with mixer, and pour classified seafood into seafood soup stock, such as crispy mussel, chewy shrimp ball and then put one foam milk whipped from clam soup stock, to enable a multi-layer taste of wonder.

Cooking time›› 1 hour.

主材料：

洋蔥	1個	1 onion
番茄	1個	1 tomato
胡蘿蔔	1條	1 carrot
韭菜	1根	1 leek
大蒜	2瓣	2 cloves of garlic
蝦	8尾	8 shrimp
蛤蜊	8個	8 clams
淡菜	8個	8 mussels
紅鰡（切成8片）	600公克	600 grams red mullet sliced into 8 segments
月桂葉		Bay leaves
Antonio Cano 0'4° O & Co橄欖油	50公克	50 grams ANTONIO CANO 0´4° O & CO brand olive oil
米	40公克	40 grams rice
洋香菜	10公克	10 grams parsley
Goccia Natura O & Co初榨橄欖油	10公克	10 grams GOCCIA NATURA O & CO brand virgin olive oil
蟹螯	4支	4 crab pincers
水	1.5公升	1.5 litres of water

新派做法材料：

Modern

O & Co番茄粉		tomato powder O & CO brand
吉利丁	8公克	8 grams gelatin

傳統 Traditional 海鮮湯做法：

1. 蔬菜切丁，入大鍋，用橄欖油炒成金黃色，放入4個蝦頭和蝦身，快炒至變紅。再放入4個蛤蜊和4個淡菜，炒到開為止。去殼後，把海鮮肉丟回鍋中；帶殼的蝦則留在鍋裡。

2. 紅鰹切片後，留4片備用。

3. 魚頭和魚骨加洋香菜，一起煮湯。只要將材料煮沸，濾掉魚骨和魚頭。把魚湯倒入蔬菜鍋中，放入米，加月桂葉，全部一起煮30分鐘。

4. 30分鐘後，放進4片紅鰹（記住，留4片備用，還有蟹螯），再煮5分鐘。將鍋裡的材料攪拌一番後，用濾篩過濾，留湯。必要的話可加適量鹽調味，但要記住，海鮮裡的蛤蜊本身就帶有鹹味。

5. 將剩下的4片紅鰹、4支蟹螯、4個蛤蜊、4個淡菜和4尾蝦，放入海鮮湯中煮滾。煮熟後，用帶孔的勺撈起，盛盤，湯則用不同的碗盛裝。上菜前，在海鮮上面滴幾滴初榨橄欖油。

Chop the vegetables and fry in a large casserole with the ANTONIO CANO 0´4º O & CO olive oil until golden. Next, add the heads of four shrimp and four whole shrimp and fry briefly until red. Now add four clams and four mussels until they open. Remove the shells but return the meat to the casserole. The shelled shrimp may also remain.

Slice the red mullet and set aside 4 pieces to use when preparing the seafood.Using the head and the fish bones, prepare a fish broth with the parsley. Simply let the ingredients come to a boil, then sieve off the fish. Now, add this broth to the casserole with the vegetables. Next, add the rice and the bay leaves and cook all together for 30 minutes. After this, add 4 of the slices of mullet (remember, set aside the remaining 4 slices saved for the seafood; this also includes the crab pincers). Cook for 5 more minutes.

Blend everything in the casserole and run through a strainer to obtain the soup. Add salt as needed, but keep in mind that the clams used for the seafood will contain sea salt.

To prepare the remaining 4 slices of mullet, crab pinchers, 4 clams, 4 mussels, and 4 shrimp, simply boil the seafood in the soup and, once cooked, remove with a slotted spoon and place on a plate. Serve the soup with separate bowl on the side. Drizzle some GOCCIA NATURA O & CO virgin olive oil over the seafood before serving the soup.

新派 Modern 海鮮湯做法：

1. 照上述傳統做法，把菜切好，魚則切片，留4片備用。

2. 將蔬菜、魚肉、米和洋香菜倒入鍋中，加水，煮湯。煮好後（約30分鐘左右），濾掉所有的固體，只留液體。

3. 把淡菜放入這鍋湯中，煮到淡菜開了，撈起來，放到一旁。接著，把4尾蝦和蟹螯丟入湯中，煮熟。再次用咖啡濾紙濾去所有的固體。

4. 小心剝好4尾蝦（不要剝壞殼），用刀身將蝦肉壓成扁扁的「餅」狀。蟹螯去殼，把蟹肉和蝦餅肉塞入蝦殼裡，捲成球狀，丟入魚湯中，煮個幾秒鐘。

5. 將淡菜的肉丟到鍋中，不倒油，慢慢乾煎，每隔30秒就翻面，煎到又乾又脆。起鍋，切丁，切得越小塊越好。

6. 把剩下的4片紅鰹和蛤蜊放入200毫升的魚湯中煮，濾掉固體後，就是一碗蛤蜊湯。

7. 把吉利丁丟入冷水裡浸泡，一化開就倒入溫熱的蛤蜊湯中。用鮮奶油發泡機把這碗加了吉利丁的蛤蜊湯打出泡來，做成蛤蜊淡菜泡後，放進冰箱冰1小時。

8. 把煎得酥脆的淡菜、蝦球和番茄粉裝盤，再倒上蛤蜊和淡菜湯打出來的奶泡。當著客人的面，把熱湯澆在海鮮上，再滴上幾滴初榨橄欖油。

Chop all the vegetables and slice the fish, setting aside 4 pieces--following the traditional method above. Next, combine the vegetables, fish, parsley, rice and bay leaves in a casserole to prepare a fish broth. When cooked (about 30 minutes), sieve off all the solids, reserving only the liquid.

Add mussels to this broth and cook until the shells open. Remove mussels from the broth and set aside. Boil 4 shrimp and the crab pincers in the broth. Sieve off the solids from the broth one more time using a coffee filter.

Peel the four shrimp (reserving the shells intact). Using the flat of a knife, crush the flesh to obtain a kind of crepe. Shell the crab pincer and stuff the meat into the shrimp shell with the crepe. Curl the shell into the shape of a ball. Boil for a few seconds in the broth.

Put the meat of the mussels in a frying pan without oil and cook slowly, turning every 30 seconds until they are dry and crispy. Remove and dice as finely as possible.

Cook the remaining 4 slices of mullet and the clams in 200 millilitres of fish broth. Sieve the solids from the liquid to obtain a clam soup. Put the gelatin in cold water and, once dissolved, add to the warm clam soup. Pour the clam soup gelatin in the cream whipper and aerate with the gas to create a clam-mullet foam. Store in the refrigerator for one hour.

Serve the crispy mussels, the shrimp ball, and the tomato powder on a plate and top with the clam and mullet foam. Pour hot broth over the seafood in front of your guests. Drizzle with some GOCCIA NATURA O & CO virgin olive oil.

Spanish "Tortilla"

西班牙蛋餅

　　這是一道所有西班牙媽媽都會做的家常菜，每個人都有自己的私房食譜，我常常開玩笑說：「西班牙有上千個 TORTILLA 食譜」，就好像每個中國媽媽都有自己的蛋炒飯配方一樣。

　　TORTILLA 是一道沒有終點線的食譜，它的基本材料是馬鈴薯、洋蔥、大蒜、雞蛋和橄欖油，簡單容易上手，幾乎人人會做，變化卻無窮，你可以隨喜好加入火腿、橄欖、青椒、番茄，多年前我曾經在北非突尼西亞吃到過加了蠍子和蜘蛛的 TORTILLA，這味食譜彈性之大可見一斑。

　　動手做 TORTILLA，最好準備一個不沾鍋的平底鍋，鍋要燒得很熱，快速將加了馬鈴薯的蛋餅兩面煎熟。摩登版則完全打破傳統蛋餅的格局，利用相同的材料，玩不一樣的烹飪遊戲，馬鈴薯變容器，塞入炸洋蔥絲，再放上一個生蛋黃，上桌用刀輕輕一劃，蛋黃如火山熔岩流敞下來，酥脆的洋蔥絲和馬鈴薯，吃起來跟傳統 TORTILLA 擁有一樣的口感，驚喜感卻大大不同。

　　This is a so-called homemade dish every Spanish mom can cook and each mom has her own private recipe. Therefore, I always said jokingly, "There are thousands of tortilla recipes in Spain." It's just like every Chinese mom has her own private recipe for fried rice with egg.

　　It's easy to make Spanish tortilla with simple ingredients of potatoes, onions, garlic, eggs and olive oil and almost anyone can do it. There are countless changes because you could add ham, olives, green pepper, tomatoes or whatever you like. Years ago when I was in Tunisia I have tasted tortilla with scorpions and spiders, so you can imagine the variety of the recipes.

　　To make tortilla, you'd better use a non-stick pan. In a real hot pan, quickly fry the potato-made tortilla until both sides are done. The modern recipe totally breaks the traditional definition for tortilla. With the same ingredients, play it differently. Use potato as a container, stuff with fried onion shreds and put a raw egg yolk on top of it. Slash egg yolk with a knife while on the table and let the yolk drip down like lava. Fried onion shreds and potato will taste as crispy as those of traditional tortilla but surprise the taste buds.

Cooking time›› 40 minutes.

主材料：

蛋	8顆	8 eggs
馬鈴薯	600公克	600 grams potato
洋蔥	100公克	100 grams onion
大蒜	20公克	20 grams garlic
荷蘭芹	10公克	10 grams parsley
橄欖油	200公克	200 grams olive oil
鹽	10公克	10 grams salt

傳統 Traditional 西班牙蛋餅做法:

1. 馬鈴薯削皮之後切片,盡量切得越薄越好。洋蔥和大蒜切碎。用橄欖油爆香洋蔥和大蒜、並將馬鈴薯片炒至變軟,將剩下的油過濾之後備用。

2. 將蛋加入炒熟的馬鈴薯,用打蛋器打約1分鐘,直到與馬鈴薯完全混合。

3. 在一只平底鍋中放入一點橄欖油,加熱至非常滾燙之後轉成小火,再加入拌勻的蛋跟馬鈴薯,稍微攪拌之後慢煎。

4. 1分鐘後,用一個盤子蓋住,將烘蛋倒扣在盤子上,之後將鍋子放回爐子上,加一些橄欖油,將翻面的薄餅煎至金黃。盛盤之後用荷蘭芹做裝飾。

Tips:
雞蛋的新鮮度非常重要!!
Freshness of eggs is very important!

Peel the potatoes and slice them as thin as you can. Mince the onion and the garlic. Fry all the sliced potatoes together slowly with the olive oil until the potato is tender. Strain the olive oil and set aside.

Now mix the eggs and the warm, cooked potatoes together, whipping for one minute until the egg is well incorporated.

Put a frying pan on the burner with some olive oil until the olive oil is very very hot, then reduce the flame to the minimum setting and add the eggs and potato. Stir a little bit and let the mixture cook slowly.

After 1 minute, using a dish as a cover, turn the tortilla out onto the plate. Next, return the fry pan to the burner and add some more olive oil, flipping the tortilla to brown the other side. Garnish with some parsley.

新派 Modern 西班牙蛋餅做法：

1. 將馬鈴薯切成 4 x 4 公分大小方塊，再用一把利刀在馬鈴薯塊中間挖出 3 x 3 公分的洞。

2. 用攝氏100度的油溫將馬鈴薯炸至熟透，要避免表面變焦脆。熟了之後用叉子試插馬鈴薯，變軟後關火取出備用。

3. 洋蔥切絲，用炸馬鈴薯剩下的油炸香，慢慢攪拌到洋蔥變金黃色。用有孔的勺子將洋蔥撈出，過濾剩下的油備用。

4. 將炸過的洋蔥放進馬鈴薯的凹洞裡，再放上一顆生的蛋黃。用電鍋蒸2分鐘，趁熱裝盤。取出馬鈴薯塊的時候動作要小心，最好只用一個鏟子。

5. 用炸洋蔥剩下的油爆香大蒜，淋在蛋黃上。撒上一點切碎的荷蘭芹做裝飾，並加一點天然海鹽調味。

Cut a large potato into a cube measuring at least 4 cm. x 4 cm. using a sharpen knife. Make a hole inside the cube that is at least 3 cm. x 3 cm.

Deep fry the potato slowly, covered with olive oil until it is tender but not toasted. The perfect temperature is 100°C. Once it is cooked and tender when forked, take off the heat and set it aside.

Slice the onion and cook it on the same olive oil used for the potato. Mix slowly until caramelized. Remove with a slotted spoon and set aside. Strain the olive oil and reserve for later use.

Stuff the potato cube with a base of caramelized onion and, over it, pour one raw egg yolk.
Heat in a rice steamer for two minutes and serve warm.

Using the olive oil reserved from the fried onions, fry some chopped garlic and drizzle the mixture over the egg yolk. Finally, add also some chopped parsley and top with some natural sea salt crystals. Be very careful when you move the cube from the rice steamer to the dish; it usually works best if you use one spatula.

Pisto (Sauted Vegetables with Egg)

什蔬燴蛋

「Pisto」是很傳統而且道地的西班牙鄉村菜，主要用料是各種地中海常見的蔬菜，如胡蘿蔔、洋蔥、番茄、南瓜、節瓜、茄子和紅黃椒，所有材料切成同樣大小之後，用橄欖油炒香，然後蓋上鍋蓋細火慢燉，讓蔬菜釋放出甘甜汁液，起鍋前再加進麵包，並打個蛋下去，就成了飽實又營養的一餐。

只要變換它的料理方式，很樸實的燉菜也可以很摩登，這裡我使用了一點巧妙手法料理雞蛋，只要利用家裡任何一個咖啡杯、一張保鮮膜，先倒一點橄欖油在保鮮膜上防止沾黏，再打一枚蛋下去，撒上巴西利、鹽和黑胡椒，轉緊收口，放在滾水中煮 4 分鐘，就可以得到一個美麗的「蛋花」。

"Pisto" is a traditional and genuine country dish in Spain, with various commonly seen vegetables in the Mediterranean used as primary ingredients such as carrot, onion, tomato, pumpkin, hairy cucumber, egg plant and bell pepper, after all ingredient were cut into same size, stir-fry with olive oil, covered and stew slowly with small fire, and enable a sweet flavor of vegetables, add on bread with one whipped egg before removing from the oven, will make a rich and nutritious meal.

As long as you change the way of cooking, a plain braised dish can be very modern, I use little skill here for egg cooking, all you need is to use any of the coffee cups, one-piece plastic film at home, put little olive oil onto the plastic film first to prevent from stickiness, whip another egg and spill with Parsley, salt and black pepper, seal the outlet tight and put into the boiling water for 4 minutes to enable a beautiful "egg flake".

Cooking time›› 20 minutes.

主材料：

茄子	100公克	100 grams aubergine (eggplant)
節瓜	100公克	100 grams zucchini
胡蘿蔔	100公克	100 grams carrot
洋蔥	100公克	100 grams onion
番茄	100公克	100 grams tomato
青椒	100公克	100 grams green pepper
紅椒	100公克	100 grams red pepper
大蒜	4瓣	4 cloves of garlic
法國麵包	50公克	50 grams French bread
Antonio Cano 0'4°O&Co橄欖油	50公克	50 gramsANTONIO CANO 0´4 º O & CO brand olive oil
紅椒粉	5公克	5 grams paprika
蛋	4個	4 eggs
鹽		salt
黑胡椒		black pepper
洋香菜末	20公克	20 grams chopped parsley

傳統 Traditional
什蔬燴蛋做法：

1. 將茄子、節瓜、胡蘿蔔、洋蔥、番茄、青椒、紅椒切丁
   （大小約1公分X3公分），放入鍋中，倒入橄欖油，小火
   慢炒到蔬菜釋出水分。
2. 加入紅椒粉和法國麵包，再煮2分鐘。然後，打蛋下去，煮
   到蛋白凝固。撒上鹽和胡椒，再起鍋，綴以洋香菜。

Cube the vegetables (1 cm. x 3) and transfer to a casserole with
the ANTONIO CANO 0´4° O & CO olive oil. Cook slowly until
the vegetables release their juices, then add the paprika, the
French bread, and cook for 2 more minutes.

Next, add the egg and mix, cooking until the egg is curdled. Add
the salt and pepper before serving. Decorate with the parsley.

新派 Modern 什蔬燴蛋做法：

1. 按照傳統做法，將茄子、節瓜、胡蘿蔔、洋蔥、番茄、青椒、紅椒切丁（大小約1公分X3公分）。但是，不要將所有的蔬菜炒在一起，倒點橄欖油，每一種菜分開來炒。炒到半熟，仍脆脆的即可。

2. 法國麵包切丁去烤，烤好了撒上紅椒粉。

3. 用一只咖啡杯和保鮮膜做「蛋花」：撕一片保鮮膜，大小約15公分X15公分，覆到咖啡杯內，貼緊杯壁。在杯裡倒點橄欖油，把蛋打進去，加點洋香菜、鹽和黑胡椒。將保鮮膜包起來，抓成花形，用束帶綁緊。用水煮沸（一般大小的蛋約煮4分鐘）。

4. 將蔬菜盛到盤中，用模具箍成圓形。蛋打開，倒在蔬菜上。用麵包丁裝飾，再滴上橄欖油。

Cube the vegetables (1 cm. x 3) as in the traditional method, but—rather than cooking all together—sauté each separately in a few rounds with a little bit of ANTONIO CANO 0´4º O & CO olive oil until half cooked and still crispy.

Cube the bread and toast it. Sprinkle the croutons with the paprika.

Using a coffee cup and plastic wrap, prepare egg flowers. Cut a sheet 15 x 15 cm. plastic wrap and line the inside of a coffee cup, pressing to adhere it to the side. Pour some ANTONIO CANO 0´4º O & CO brand olive oil into the bottom, then crack the egg into the cup. Next, add some parsley, salt, and black pepper. Close the plastic wrap bag, shaping the egg into a flower. Tie it securely with a zip tie. Cook in boiling water (about 4 minutes for a medium sized egg).

Place the vegetables in the centre of a plate using a ring to mould into a round shape, open the egg and settle it over the vegetables. Garnish with the croutons and drizzle with olive oil.

牛肝菌黑松露炒蛋

　　經典的松露炒蛋食譜來自法國南部，相傳這道菜的研發完全是一個意外，人們將新採的松露跟雞蛋擺在一起，生雞蛋上的毛細孔吸收了松露強烈的氣味，端出來的炒蛋特別有味道。純就味覺來說，雞蛋的味道跟松露非常match，難怪許多法國名廚都公認雞蛋是表現松露的最佳拍檔。

　　西班牙的西北部樹林裡也產有黑松露，當地人最常見的吃法正是炒蛋，偶爾也會加入氣味濃郁的牛肝菌，料理手法相當簡單，但矜貴的松露和牛肝菌的口感，讓這道菜加分不少。

　　在新式做法裡，我把原本一鍋炒的蛋白和蛋黃分開，利用打發的蛋白，為松露蘑菇湯營造出如同雲朵般的輕柔口感，一樣的食材，馬上就有了不一樣的生命。

This classic scrambled egg with truffle recipe comes from south of France, it is heard that this course is purely invented out of an accident, people put eggs beside newly-harvested truffle, in which the pore of raw eggs absorbs strong smell of truffle and enables a special flavor to the scrambled egg. As far as sense of taste is concerned, eggs' smell makes a lovely match to truffle, no wonder egg is generally acknowledged as best partner to truffle by many noted French chefs.

There are black truffles in the northern forest of Spain, scrambled egg is the most common way to eat and will add on bolteus with strong smell sometimes, which is a very simple way of cooking, however, the precious truffle and taste of bolteus have even enabled a better taste to this course.

In new way of cooking, where I separate egg white from yolk, and use whipped egg white to create a soft-like-a-cloud taste for truffle mushroom soup, a different life with same ingredients.

Cooking time›› 10 minutes.

主材料：

蛋	4個
牛肝菌	200公克
O & Co黑松露	2公克
鹽	5公克
黑胡椒	5公克
Antonio Cano 0'4° O&Co橄欖油	10公克

新派做法材料：

蛋	4個
O & Co松露蘑菇湯	1罐
Antonio Cano 0'4° O&Co橄欖油	10公克

4 eggs
200 grams porcini mushroom (Boletus Edulis)
2 grams O & CO brand black truffle
 (tuber melanosporum)
5 grams salt
5 grams black pepper
10 grams ANTONIO CANO 0´4° O&CO brand
 olive oil

Modern
4 eggs
1 jar O & CO brand Crema de Funghi Porcinni
 con Tartufo (cream of mushroom with truffle)
10 grams ANTONIO CANO 0´4° O & CO brand
 olive oil

Scrambled Egg with Boletus
Edulis and Black Truffle
D.I.Y (do it yourself)

傳統 Traditional
牛肝菌黑松露炒蛋做法：

1.牛肝菌切塊，用橄欖油加鹽和黑胡椒炒香。

2.牛肝菌一軟，倒入打好的蛋，以文火煎
  到呈淺黃色。

3.蛋煎好，撒上碎松露，立即食用。

Cut the porcini mushrooms into square
pieces and sauté with the ANTONIO
CANO 0´4º O & CO olive oil, salt, and
black pepper.

Once the mushrooms are soft, add the
mixed eggs on low heat until creamy.
When the scrambled eggs are
done, grate the truffle over
the top and serve
immediately.

新派 Modern 黑松露蘑菇佐蛋白慕斯做法：

1. 打蛋，將蛋黃和蛋白分開。
2. 用打蛋器把蛋白打到堅挺發泡。
3. 將松露蘑菇湯倒入鍋中加熱，千萬不要煮沸。舀一大匙松露蘑菇湯，倒入盤中。將蛋黃放在湯上，再倒一大匙打好的蛋白泡。
4. 用大火熱橄欖油，淋到桌上的盤子裡，請客人自行拌料，就可嚐到混合的美味。

Crack the eggs and separate the yolks from the whites. With a mixer, whisk the egg whites until they are stiff and form peaks.

In a pan, heat the O & CO Crema de Funghi Porcinni con Tartufo, never letting it boil.

In the centre of a dish, spoon one tablespoon of the O & CO Crema de Funghi Porcinni con Tartufo. Settle the egg yolk over the mushroom cream and a top with a tablespoon of the whisked egg white.

Heat the ANTONIO CANO 0´4º O & CO olive oil on high and then drizzle over the dish at the table. Have your guests mix the ingredients for a fabulous taste combination.

Crab Meat Donostiarra Style

多諾斯蒂亞式蟹肉

　　這是從傳統 Tapas 演變而來的一道菜，傳統的做法厚味重口，正好下酒，配著雪莉酒一起吃，味道剛剛好。

　　但是台灣客人的口味普遍偏輕，我在「EL　TORO」餐廳推出這道菜時，有些客人嫌它味道重，後來我發現只要在菜裡多加一項秘密武器，就可以大大改變這道傳統食譜的味道和觀感，這個神奇的秘密武器，是很多人怎麼想都想不到的——白巧克力！

　　鮮鹹的蟹肉，淋上甜甜的白巧克力醬，鹹味消減了，反而帶出蟹肉的鮮甜美味，這是烹飪上 1+1 > 2 的神奇魔術，至於味道有多 Match？

　　你試了就知道！

　　A dish that evolves from traditional Tapas, featuring strong smell and taste, conventional crab chowder is good to serve with wine, which is just perfect taste if serve with Sherry wine.

　　However, due to a preference of light flavor by Taiwanese customers, its strong flavor is not favored by some clients when I first released this course at "EL TORO", I soon realized that the taste and perception on this traditional recipe will largely be changed once one more secret ingredient is added on the dish, which is something beyond your expectation - white chocolate!

　　Fresh and salty crab chowder sprinkled with sweet white chocolate will help eliminate the salty taste and enables a sweet and fresh taste of crab meat, which is a 1+1>2 magic for culinary art, but how fit it is? You will know as soon as you take it a try!

Cooking time›› 2 hours.

主材料：

螃蟹（800公克）	1隻	1 crab (800 grams)
洋蔥	1個	1 onion
胡蘿蔔	1條	1 carrot
大蒜	2瓣	2 cloves of garlic
麵包心	20公克	20 grams bread crumbs
奶油	100公克	100 grams butter
Antonio Cano 0′4°O&Co橄欖油	25公克	25 grams ANTONIO CANO 0´4º O & CO olive oil
月桂葉		bay leaves
番茄醬	250公克	250 grams tomato sauce

新派做法材料：

Modern		
白巧克力	50公克	50 grams white chocolate
牛奶	25公克	25 grams milk

傳統 Traditional
多諾斯蒂
亞式蟹肉做法：

1. 胡蘿蔔和洋蔥切塊，切得越小塊越好，加奶油和橄欖油攪拌，放在鍋裏慢火炒到金黃色。

2. 用另一只鍋裝滿5公升的水，加入50公克左右的海鹽和一片肉桂葉，煮蟹。每100公克的螃蟹要煮1分鐘，800公克的蟹大約要煮8分鐘左右。

3. 螃蟹去殼（留住殼裡的汁），挑出蟹肉拌入蔬菜丁和麵包心，煮到收乾湯汁，其間要不斷地攪拌，以免黏鍋。

4. 沸騰後加入幾片月桂葉和番茄醬，攪拌後再煮到開，即可食用。

Cut all the vegetables as small as possible and combine in a casserole with the butter and the ANTONIO CANO 0´4º O & CO olive oil; cook it slowly until golden.

Meanwhile, cook crab in a casserole with 5 litres of water and about 50 grams of sea salt. Add one bay leave and cook the crab 1 minute for each 100 grams. In this case, cooking time will be about 8 minutes.

Separate all the crab meat from the shell with patience, retaining all the leftover juice inside the shell. Next, mix the crab meat with the vegetables and the bread crumbs and cook it until boiling, MIXING ALL THE TIME. This prevents the mixture from becoming sticky. Once it is boiling, add a few bay leaves and the tomato sauce; mix everything and boil again. Now it is ready to serve!

新派 Modern 多諾斯蒂
亞式蟹肉做法：

這新派做法只要多加一種材料就好。但是，根據個人所見，光是這項變化就大大地改變這道傳統食譜。這項特色就是白巧克力！

在白巧克力裝入玻璃杯中，倒入牛奶，放進微波爐加熱融化。在盤裡放上做好的蟹肉，旁邊搭配一杯杯熱熱、稠稠的巧克力。

要吃的時候，把巧克力倒在蟹肉上。鹹鹹甜甜的滋味，令人驚奇，肯定叫你喜愛！

For this recipe, simply add one more ingredient. But this one change, in my opinion, significantly modifies the classic recipe.

The special touch is the white chocolate! Place the sweet chocolate in a glass with the milk and melt in the microwave. Put a dollop of crab in a bowl and serve the warm, creamy liquid in small shot glasses on the side.

When ready to eat, pour chocolate over the crab. With the surprising mixture of sweet and salty, you'll be sure to enjoy this taste sensation!

加利西亞章魚盤

被大西洋環抱的加利西亞自治區（Galicia），位於伊比利亞半島的西北部，素來就以種類繁多的海鮮著稱，牡蠣、龍蝦、扇貝、藤壺、章魚、烏賊…，不但是各色菜餚的主角，也造就各種用海鮮做的下酒菜（Tapas）。

章魚盤正是該區最經典的 Tapas 之一，在市集工作的人們休息的時候聚在一起，把手邊新鮮的章魚用滾水煮熟，切片淋點橄欖油、灑上紅椒粉和海鹽，就這麼邊吃邊喝邊聊了起來。

傳統的章魚盤就這麼簡單到令人不可置信，但因為夠新鮮，怎麼做都好吃。新派做法中我只是換了一些呈現方式，用吉利丁做出一個容器，裡面放入紅椒粉、橄欖油和海鹽，當炙熱的章魚碰到吉利丁凍，逐漸被熱氣融化流出高湯，不但激出紅椒粉和橄欖油的香氣，也讓這道菜產生豐富的表情變化。

Surrounded by the Atlantic Ocean, Galicia is an autonomous community in northwestern Iberian Peninsula. This region has long been famous for its various kinds of seafood, to name but just a few: oysters, lobster, sea fans, barnacles, octopus and squid. Not only is seafood used to make main dishes, but it is also used to prepare Tapas.

Octopus a feira is one of the typical tapas. People who work at fairs gather together during break time. They boil fresh octopus at hand, slice it, then pour some olive oil over, and sprinkle some paprika and sea salt. Afterwards, they talk while they eat and drink.

Although it's such an incredibly simple dish, it's tasty because of its freshness. I just present it in a different way: use gelatin jelly as a container to hold paprika, olive oil and sea salt. When the steamy hot octopus meet the icy cold jelly, the jelly soon melt and the stock flows out, thus spurring the aroma of paprika and olive oil and bring a new look to an old dish.

Cooking time›› 1 hour.

主材料：

大章魚（3公斤）	1隻	1 big octopus (3 kg)
洋蔥	1個	1 onion
月桂葉		bay leaves
鹽	100公克	100 grams salt
紅椒粉	15公克	15 grams paprika
初榨橄欖油	100公克	100 grams virgin olive oil
O & Co葡萄牙海鹽	適量	100 grams O & CO brand FLEUR DE SEL (Portugal)

新派做法材料： **Modern**

吉利丁	16公克	16 grams gelatin
烤好的法國麵包	50公克	50 grams pre-cooked French bread

Octopus À Feira Style

Spécialité à base d'
Huile d'Olive
Vierge Extra et
au Piment
Extra Virgin
Olive Oil
specialty
with Chili Pepper
OLIVIERS & CO.® - 04300 MANE - FRANCE.
50 ml ℮ 1.7 fl.oz

O&CO.

傳統 Traditional 加利西亞章魚盤做法：

1. 章魚洗淨。
2. 倒入8公升的滾水到大鍋中，加鹽，放入剝好的洋蔥，再下章魚，以中火煮1小時。
3. 撈出章魚，切片，加點紅椒粉、初榨橄欖油和一小撮海鹽，即可食用。

Clean the octopus. Pour 8 litters of boiling water into a large casserole. Add the salt, the bay leaves, and the peeled onion. Next, add the octopus and cook over medium heat for 1 hour.

Slice the octopus and serve with some paprika, virgin olive oil, and a sprinkling of O & CO FLEUR DE SEL.

新派 Modern 加利西亞章魚盤做法：

1. 用上述傳統手法煮章魚，煮好後鍋裡留下500毫升的水。
2. 吉利丁放進冷水裡泡軟，倒入煮章魚的溫水中，再將之倒入4X3公分大小的模具中，放入冰箱，冰1小時。
3. 拿掉模具，用一只咖啡匙和一把鋒利小刀，在吉利丁中央挖個洞，底部至少要留0.5公分的厚度，倒入紅椒粉、初榨橄欖油和海鹽，再收進冰箱裡。
4. 麵包放入果汁機中，加入10公克煮章魚的高湯和5公克紅椒粉打成「糊」，攤在烘焙紙上，越薄越好。
5. 將麵包糊送進預熱至攝氏150度的烤箱中，烤到酥脆。
6. 章魚鬚橫切成片。拿靠近頭部、比較厚的章魚鬚，用熱鍋炙烤（比較薄的章魚鬚可收留著做沙拉）。
7. 將熱騰騰的章魚擺到盤中，配上烤出來的紅椒粉章魚餅乾。
8. 在餐桌上，當著客人的面，將吉利丁凍放到熱騰騰的章魚上，看著吉利丁被章魚的熱氣融化，逼出橄欖油、紅椒粉和加味海鹽的香氣。

Cook the octopus following the same method as the traditional recipe above, but retain 510 ml. of the water from the casserole when done.

Mix the gelatine into the cold water until soft, then add to the 500 ml. of warm octopus water. Pour the mixture into a round 4 cm by 3 cm mould and transfer to the refrigerator. Chill for 1 hour.
Remove the gelatine mould and, in the centre, use a coffee spoon and a small sharpen knife to hollow out a cup. The bottom must be at least 0.5 cm thick. In the well, add the paprika, the virgin olive oil and the O & CO FLEUR DE SEL. Return to the refrigerator.

Put the bread in a blender and chop, adding 10 grams of the octopus water and 5 grams of paprika. This will create a paste that you will spread over oven paper, rolling as thin as possible. Bake the dough at 150ºC until crispy.

Slice the octopus tentacles width-wise, using the thicker pieces closer to the head. Sear the rounds in a very hot pan. (Retain the thinner tentacles to use in a fantastic salad).

Serve the octopus very hot in the centre of the dish with the paprika cookie accompanying. At the table before your guests, settle the gelatine over the hot seafood and watch with delight as the heat of the octopus melts the gelatine-- releasing the perfume of the olive oil, paprika and flavoured salt.

Squid´s Ink Rice
Cooked On Paella

西班牙墨汁烏賊燉飯

　　黑色的「西班牙墨汁烏賊燉飯」原本並未放在「 EL TORO 」的日常菜單上，一位美食部落客在餐廳吃過後非常喜歡，在他的網誌上做了推荐，此後好長一段時間，餐廳不斷有人上門指名要點這道黑色的燉飯。

　　跟西班牙海鮮飯一樣，這也是一道經典食譜，但跟金黃色的海鮮飯最大不同點，在於墨汁燉飯中沒有加番紅花，卻加了黑黑的墨魚汁，正統的墨汁燉飯只用墨魚或小烏賊烹製，後來花樣逐漸多變，加進燉飯的食材也越形豐富。

　　在摩登版的做法中，我打破傳統西班牙海鮮飯的模樣，改將燉飯塞入墨魚的身體裡，讓傳統菜式有了全然不同的外觀，但是它的口感和食材卻是大同小異的。墨汁燉飯非常美味，唯一要注意的是，吃完這道飯後，切記不要立刻露齒微笑，免得別人看到你的———滿嘴大黑牙！

　　In the beginning Squid´s Ink Rice Cooked on Paella was not listed on the daily menu of El Toro. It so happened that a food blogger enjoyed this dish so much after dining at our restaurant that he recommended it on his web log. Ever since then, people come and ask for this black paella.

　　Just like Spanish seafood paella, it's a classic dish. The biggest difference between these two is that squid's ink, instead of saffron, is used in the black paella. Typical squid's ink cooked paella just uses squid or cuttlefish and it is only later that we add more and more ingredients and make a lot of changes.

　　By stuffing the ink-cooked rice into the squid body, I have broken the rule of the presentation of Spanish paella. Although I give this traditional dish a totally different look, the taste and ingredients are almost the same. Enjoy this delicious dish but remember — don't grin after eating so that people won't see your black teeth!

Cooking time›› 45 minutes.

主材料：

墨魚	400公克	400 grams squid
洋蔥	50公克	50 grams onion
紅椒	20公克	20 grams red pepper
青椒	20公克	20 grams green pepper
大蒜	10公克	10 grams garlic
米	200公克	200 grams rice
新鮮墨魚汁	20公克	20 grams fresh squid ink
魚湯	180公克	180 grams fish broth
鹽		salt
Antonio Cano0' 4° O & Co橄欖油	25公克	25 grams ANTONIO CANO 0´4° O & CO brand olive oil
初榨橄欖油		virgin olive oil
洋香菜		parsley

新派做法材料：

		Modern
帕瑪乾酪	50公克	50 grams Parmigiano Reggiano
魚湯	50公克	50 grams fish broth
洋蔥	100公克	100 grams onion

傳統 Traditional
墨汁烏賊燉飯做法：

1. 輕輕搓洗墨魚，去除內臟，留下墨魚身連墨魚鬚。墨魚身切片，約2公分寬大小。一半烹飪用，一半留做擺盤裝飾。

2. 洋蔥、紅椒和青椒切丁，備用。再將墨魚汁倒入魚湯中煮開，拌成深黑色的醬。

3. 在海鮮飯專用鍋中倒入橄欖油，放下蔬菜丁炒香，倒入墨魚片，炒熟。再倒入米炒30秒，拌勻後，倒入黑色墨魚高湯，轉中小火煮10分鐘，讓米飯吸飽醬汁，熬煮時千萬不要刮到鍋底，否則會攪到黏稠的那層（聞起來可能會有點焦味，沒有關係）。

4. 10分鐘後起鍋，移進烤箱，續烤6分鐘後，將留下來的墨魚鬚和墨魚片擺到飯上，再烤1分鐘，起鍋後淋上初榨橄欖油再出菜，並用洋香菜葉做裝飾。

Clean the squid gently, removing all the internal organs, reserving the body with tentacles.

Cut the body into 2 cm. wide slices. Half of the slices will be for the principle recipe and half are for the garnish. Dice the vegetables and set aside. Cook the fresh ink with the fish broth, blending to obtain a deep black sauce.

In a paella pan, heat the vegetables with the ANTONIO CANO 0´4° O & CO brand olive oil and fry until golden, then add the squid and cook until toasted. Then add the rice and fry for 30 seconds, mixing everything. Next, add the black sauce and cook slowly for 10 minutes, allowing the rice to absorb the sauce. NEVER scrape the bottom of the paella pan since it will disturb a sticky layer. (It may smell a little burned, but that is ok). After 10 minutes, place the pan in the oven for 6 more minutes, topping the rice with the reserved tentacles and sliced squid.

Before serving, wait one minute and add the virgin olive oil. Garnish with some parsley leaves.

新派 Modern 墨汁烏賊燉飯做法：

1. 墨魚洗淨，分成三部分：封閉的圓筒狀墨魚身用來填料；
   墨魚鰭跟飯一起煮；墨魚鬚用來裝飾。

2. 蔬菜切塊，炒熟但不失脆感。

3. 將洋蔥切成薄片，放進橄欖油中慢慢加熱，等洋蔥變成焦
   糖色，濾出備用，剩下的油則用來燉飯。

4. 熱鍋，不放油先將墨魚煎熟。接著再炙墨魚鬚，注意火
   候，不要烤到捲起來。

5. 先倒少許橄欖油，煎香墨魚鰭，再倒入米，並再淋點橄欖
   油，然後拌入黑色的墨魚醬汁轉中小火慢煮。等湯汁開始
   收乾，再倒入50CC魚湯，讓飯更柔滑。煮時要不停攪拌。
   等飯一軟先倒入一半橄欖油，並加進20公克的巴馬乾酪，
   拌勻。

6. 洋香菜末拌入剩下那一半的橄欖油中，拌成綠色橄欖油。

7. 將料一一塞入圓筒狀的墨魚身內，從焦糖色的洋蔥、燉飯
   到炒好的蔬菜，最後再塞入更多的飯。用一條墨魚鬚封
   管，再塞點巴馬乾酪，淋上綠色的橄欖油。

Wash and divide the squid into three sections to create a closed cylinder with the body cone for stuffing. Save the fin portion of the body to cook with the rice, and tentacles to decorate.

Cut the vegetables into squares, then sauté until cooked but still crispy. Cut the additional onion into thin slices and cook slowly in the ANTONIO CANO 0´4º O & CO brand olive oil until caramelized. Next sieve the oil off and set aside for cooking the risotto.

In a very hot fry pan, sear the squid without oil until toasted. Next, sear the tentacles, trying to prevent them from curling in the pan.

Cook the remaining pieces of the fin portion of the body in a pan with some ANTONIO CANO 0´4º O & CO olive oil. Add the rice and some olive oil; next, stir in the black sauce and cook slowly. As the rice is beginning to dry, add the 50 grams of fish broth to make it more creamy. Mix continuously. Once the rice is soft, add half of the virgin olive oil and 20 grams of parmesan, mixing to incorporate.

Using the remaining half of virgin olive oil, blend it with the parsley to create a green olive oil.
Stuff the squid cylinder with layers of ingredients, beginning with the caramelized onion, some risotto, sautéed vegetables, and—finally--some more risotto Close the tube using a tentacle and top with some parmesan. Drizzle with the green virgin olive oil.

Spanish Paella

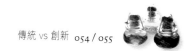

西班牙海鮮鍋巴飯

　　我的餐廳「EL TORO」剛開幕那陣子，每天都有客人跑來指名要吃「Paella」，我第一次領教到西班牙海鮮飯在台灣的高知名度。起源於西班牙東南部瓦倫西亞省的「Paella」，堪稱西班牙「國菜」，漁民利用當地盛產的魚鮮和稻米，加入香料混煮成飯，烹飪手法雖然簡單，但由於混合了海鮮滋味，煮出來的飯異常鮮美。

　　Paella 這個字原指一種用來烹調西班牙飯的平底鍋，是做海鮮飯不可或缺的元素，尤其米飯在鍋底結出一層焦脆的鍋巴，是很多老饕吃海鮮飯時的必搶目標。在新式做法中，我直接將海鮮串成串燒，再把烤過的串燒放在炒成鍋巴的飯上，配檸檬香醋醬和熱好的魚高湯，真是人生一大享受！

There were clients came to my restaurant and ordered "Paella" during the opening period, which enabled an understanding on high awareness of seafood Paella rice in Taiwan to me. Originated from southeast of Valencia in Spain, Paella, proclaimed as "national cuisine", where fishers used local fresh fishes and rice to mix with spice until they are fully cooked. A simple culinary method, but exceptionally tasty for rice mixed with seafood flavor.

Paella, refers to a pan that cooks Spanish rice originally, is now an element that is requisite for making seafood rice, with a layer of crispy rice formed underneath the bottom of pan especially, is must-to-eat target for gluttons while eating the seafood rice. In new way of cooking, I directly string up the seafood and put the roasted seafood string on the crispy rice, and serve with lemon ginger sauce and the heated fish soup stock, which is really a joy to life.

Cooking time›› 20 minutes.

主材料：

材料	份量	Ingredients
米	100公克	100 grams rice
紅椒	1個	1 red pepper
青椒	1個	1 green pepper
洋蔥	1個	1 onion
番茄	1個	1 tomato
大蒜	2瓣	2 garlic cloves
番紅花	1公克	1 gram saffron
匈牙利紅椒粉	1公克	1 gram paprika
魚湯	200公克	200 grams of fish broth
Antonio Cano 0'4° O & Co橄欖油	25公克	25 grams of ANTONIO CANO 0´4° O & CO brand olive oil
蝦（1公斤約10到14尾）	4尾	4 shrimp (10 to 14 shrimp per kilo)
活的淡菜	4個	4 fresh mussels
大蛤蜊	4個	4 large clams
雞肉	40公克	40 grams chicken
西班牙香腸	40公克	40 grams Spanish sausage (chorizo)
白肉魚混鱸魚肉	80公克	80 grams white fish with flavourful meat (sea bass)
墨魚	1隻	1 squid
檸檬	1個	1 lemon
初榨橄欖油	25公克	25 grams virgin olive oil

傳統 Traditional 西班牙海鮮鍋巴飯做法：

1. 大蒜、青椒、紅椒、番茄和洋蔥切好，放進煮海鮮飯專用的鍋中，
   倒入油，微微炒過。倒入米，翻炒30秒，再加入番紅花和紅椒粉，
   再次翻炒。
2. 陸續加入西班牙香腸、雞肉和切片的墨魚，再煮30秒。
3. 倒入魚湯，不要再攪拌，以中火煮10分鐘後，加入蛤蜊、魚肉、淡
   菜和蝦，煮到米飯吸飽了湯汁（約再5分鐘，總共15分鐘）。靜置2
   分鐘，上桌前用點檸檬角和初榨橄欖油裝飾。

Using a paella pan, cook the chopped garlic, green pepper, red pepper, tomato, and onion in the ANTONIO CANO 0´4° O & CO olive oil until slightly fried. Add the rice and stir, frying for 30 seconds.

Next, add the saffron and the paprika. Stir one more time, then add the chorizo, the chicken, and the sliced squid. Cook for 30 more seconds and add the fish broth. After adding the broth, do not mix anymore.

Simply let it cook over medium heat for ten minutes. Next, add clams, shrimp, fish, and mussels--cooking until the rice absorbs the water (about 5 more minutes, or 15 minutes total). Let the paella sit for two minutes before serving and garnish with some lemon wedges and virgin olive oil.

Cook the rice in 1 litre of water for 35 minutes. When cooked, sieve and rinse the rice. Spread the rice over oven paper and let it dry in the oven for 3 hours at 80ºC. Once the rice is dry, store in a dry place.

Cut the vegetables and set aside.

Next, using the juice of the lemon, prepare a vinaigrette with the virgin olive oil.

Steam open the clams in a pan with some water. When ready, remove the shell and transfer the clams to the refrigerator. Using the remaining hot clam juice, steam open the mussels, remove the shell, and transfer to the refrigerator. Next, to this soup, add the fish broth and cook until the volume is reduced by 30%, creating a concentrated stock. Allow the hot soup to become warm, then add the saffron and the paprika, stir together and set aside.

Slice the squid and sear slightly to prepare in advance. Set aside.

Using bamboo skewers, prepare mixed brochettes of fried squid, shrimp, the clams, cubed chicken, cubed chorizo, the mussels and the fish.

Heat the ANTONIO CANO 0´4º O & CO olive oil until it is very hot and smokes, then deep fry the dry rice to inflate it. Remove the rice from the pan and transfer to a bowl lined with paper napkins to remove the excess oil.

Sear the brochettes in a pan with oil and some salt. Heat the fish stock until it´s almost boiling.

Add the rice to the centre of a plate. Settle the brochette over the re-heated rice and serve immediately.

For a garnish, serve the lemon vinaigrette and the hot fish broth (also known as a GUO- BA sauce). Mix everything and enjoy!

新派 Modern 西班牙海鮮鍋巴飯做法：

1. 用1公升水煮飯，煮35分鐘。
2. 將煮好的飯，鋪在烘焙紙上。放進預熱至攝氏80度的烤箱中，烤3小時。烤完，把飯收到乾燥處。
3. 所有食材切好備用。
4. 用檸檬汁和初榨橄欖油調成油醋醬。
5. 鍋裡裝少許水，把蛤蜊煮開，去殼後放入冰箱。利用餘下的蛤蜊湯汁把淡菜煮開，一樣去殼後放入冰箱。
6. 把魚湯倒入這鍋「湯」裡加熱濃縮到剩7成的湯汁。放涼後加入番紅花和紅椒粉，攪拌均勻。
7. 墨魚切片、微炙，備用。
8. 用竹籤把烤墨魚、蛤蜊、蝦、雞丁、香腸丁、淡菜和魚肉等串成串燒。
9. 鍋裡倒入橄欖油加熱，到冒煙了再來炒飯，把飯炒成鍋巴。炒好後起鍋盛到碗中，碗裡要先鋪餐巾紙吸油。
10. 鍋裡倒點油，撒少許鹽，烤海鮮串燒。另外把魚高湯加熱，熱到快要滾沸。
11. 飯重新熱過，盛到盤中，擺上一串串的串燒，立即上菜。食用時可以用檸檬香醋醬和熱好的魚高湯，搭配串燒享用。

燴煮紅鰹鮮蝦

　　傳統的燴煮海鮮鍋，可以任選手邊最容易買到的魚鮮材料，簡單用洋蔥、大蒜炒香，配著馬鈴薯、青椒、紅椒和番紅花粉、紅椒粉一起燴煮，煮到蔬菜軟了，所有魚蝦的美味都融入湯汁裡，端上桌，配著麵包、白酒或雪莉酒，就是豐富又飽足的一餐。

　　同樣的主材料，如何讓它有截然不同的感官？我的方法是充分利用蝦頭裡的精華來煉成「蝦油膏」，一部分拌海鹽做成紅色的調味料，一部分拿來做餅乾，這樣既能讓一道菜具有裝飾性，又能完整保留蝦膏之精華，最後再加上南洋酸橙香料葉，味覺上也多了耳目一新的感覺。

Traditional braised seafood pot chooses the most easy-to-get fish ingredient, simply stir-fry the onion and garlic, and braise along with potato, green pepper, red chili, saffron powder, chili powder until all vegetables are fully cooked, taste of all fishes and shrimps will all go into the soup, a plentiful and complete meal served on the table by serving with bread, white wine or Sherry wine.

How to make a separate feeling to sense organ with same primary ingredients? My way of doing this is to fully extract the shrimp head as "shrimp paste", with some to mix with sea salt as red seasonings, while some to make cookies, which makes a decorative dish but also completely preserve the extract of shrimp paste, and lastly, add on sour range spice leave of Southeast China, which brings everything new and fresh feeling to the sense of taste.

Cooking time ›› 30 minutes.

主材料：

紅鰹（連皮帶鱗）	800公克	800 grams red mullet (with the skin and scales on)
馬鈴薯	2大個	2 large potatoes
青椒	1個	1 green pepper
紅椒	1個	1 red pepper
洋蔥	1個	1 onion
大蒜	2瓣	2 cloves garlic
蝦	12尾	12 shrimp
Antonio Cano 0'4° O & Co橄欖油	50公克	50 grams ANTONIO CANO 0´4° O & CO brand olive oil
番紅花粉	1公克	1 gram saffron powder
匈牙利紅椒粉	2公克	2 grams paprika
洋香菜		Parsley

新派做法材料：

Modern		
法國長棍麵包	20公克	20 grams French baguette
紫薯	4個	4 purple potatoes
Goccia Natura初榨橄欖油		GOCCIA NATURA O & CO brand virgin olive oil
酸橙葉（藍象牌）		BLUE ELEPHANT brand kaffir lime leaves
鹽	5公克	5 grams salt

Red Mullet and Shrimps
Hot Pot "Caldereta"

傳統 Traditional

燴煮紅鰹鮮蝦做法：

1. 紅鰹去骨留肉，魚頭和魚骨加500毫升水熬成湯。

2. 蝦摘頭，倒點橄欖油在鍋中，將蝦頭煎熟，盛起放著備用。

3. 蔬菜切好，馬鈴薯削皮切丁，倒進剛才煎蝦子的鍋中，很快地煮一下，加入紅椒粉和番紅花粉。倒入魚湯，煮到馬鈴薯變軟。

4. 放入剝好的蝦身和切片的魚，把所有的材料煮滾。起鍋後，放點洋香菜末點綴。

Fillet the fish and reserve the meat. Using the head and bone, prepare a fish broth with 500 millilitres of water. Remove the shrimp heads and fry the heads in a casserole with ANTONIO CANO 0´4º O & CO olive oil until toasted. Remove the heads and set aside. In the same casserole, add the chopped vegetables and the peeled, cubed potato.

Cook briefly and add the paprika and saffron. Next, add the fish broth and cook until the potato is soft.

Next, add the peeled shrimp bodies and the sliced fish. Bring all the ingredients to a boil, then remove from the heat and serve using with some chopped parsley as decoration.

新派 Modern 燴煮紅鰹鮮蝦做法：

1. 紅鰹去骨，每片都切成兩半。魚頭和魚骨用500毫升的水，加蔬菜、番紅粉和紅椒粉煮湯。湯沸後，濾渣，熬成20毫升的高湯。

2. 利用蝦頭來做「蝦油膏」：頭去殼，只留裡面軟軟的部分，用橄欖油以文火慢慢煎（約煎1小時左右）。濾掉渣渣，只留蝦油，倒入果汁機中，加入魚湯調到最高速拌打。再過濾一次後，放入冰箱，凍成蝦油膏。凍好以後，取幾匙，拌點海鹽，做成紅色調味料。

3. 剩下的蝦油膏拿來做餅乾：將它拌入法國麵包，做成麵糰，攤在烘焙紙上再蓋上一張烘焙紙。把麵糰送進預熱至攝氏150度的烤箱中（烤6分鐘左右）烤到酥脆。

4. 紫薯放入微波爐煮軟，丟進果汁機裡，倒入橄欖油後，打成泥。

5. 熱鍋，煎魚。將魚肉放進加溫的烤箱中，只要幾分鐘，讓魚肉變成半透明就好。魚肉放到紫薯上，淋上幾滴初榨橄欖油，用蝦油膏在盤子上「畫」幾筆。最後，放上餅乾、撒點紅色的海鹽和藍象牌的酸橙葉，創造感官享受。

Cut each fish fillet into two pieces. Using the head and bones, prepare a broth with 500 millilitres of water, adding vegetables, saffron and paprika. Once it boils, sieve it and reduce the stock to 20 millilitres.

With the shrimp heads, prepare an essence. Remove the shells from the heads and use only the soft organs inside. Add the head organs to the ANTONIO CANO 0´4º O & CO olive oil and cook on low heat very slowly (approximately one hour). Next, sieve the head organs off and, using only the shrimp oil, blend in a mixer--adding the fish broth--at maximum speed. Sieve again and store in the refrigerator. After it has chilled, blend a few teaspoons of the shrimp essence with some sea salt to obtain a red coloured condiment.

With half of this shrimp essence, prepare a cookie. Blend the essence with the bread. Then spread the dough between two sheets of oven paper. Cook at 150 ºC until crispy (around 6 min).

Next, cook the potatoes in the microwave until tender. Afterwards, blend with GOCCIA NATURA O & CO olive oil to get a purée.

Seal the fish in a very hot pan and transfer the fillets to a hot oven for just a few minutes until the flesh is translucent. Serve the fish over the potato. Drizzle with some GOCCIA NATURA O & CO virgin olive oil and paint the dish with some shrimp essence. Finally, decorate with the cookie, some red salt, and some BLUE ELEPHANT Kaffir lime leaves to create a delight for senses!

松子雞肉卷佐摩典納甜醋

　　傳統的「什蔬松子燉雞肉」來自西班牙中部地區，最大特色是使用多量的紅酒醋或雪莉酒醋來燉煮雞肉和蔬菜，為了克服雞胸肉燉煮後容易乾柴的缺點，雞胸肉下鍋前先用炙熱的平底鍋快速煎封住表面，以留住肉汁。

　　這是一道很飽實的鄉村菜，耶誕節前後經常出現在西班牙人的餐桌上。由於用料平實，家常燉煮花不到新台幣 200 元就可以餵飽一家四口。我利用一樣的材料，但小小改變它們的烹調方式，例如將雞胸肉以蒸代燉；以焦糖化的摩典納酒醋，營造紅酒醋和炙燒雞肉的香酥口感，再淋上碎紙花一般的胡蘿蔔油，質樸的老菜煥發出令人眼睛一亮的新意！

　　Traditional chicken stew with mixed vegetables and pine nuts is a dish from central Spain. The main characteristic of this dish is to use plenty of red wine vinegar or sherry vinegar to cook chicken and vegetables. To avoid the meat of chicken being too dry after braising and to keep the juice, sear chicken breast in a hot pan before braising.

　　This is a solid country dish. It often appears on the dining table in a Spanish house during Christmas. On less than two hundred NT dollars you can buy and cook those solid ingredients to feed a family of four. With the same ingredients I've made just a few changes in cooking. For example, steam the chicken breast instead of braising, use caramelized balsamic vinegar from Modena to enhance the crispness of seared chicken, and, finally, pour carrot oil on top of the dish. Voila, a simple old dish with new ideas could be impressive.

Cooking time›› 30 minutes.

主材料：

雞胸肉	400公克	400 grams chicken breast
松子	40公克	40 grams pine nuts
杏仁	20公克	20 grams almonds
馬鈴薯	200公克	200 grams potatoes
胡蘿蔔	1根	1 carrot
初榨橄欖油	50公克	50 grams virgin olive oil
荷蘭芹	適量	parsley
鹽	適量	salt
黑胡椒	適量	black pepper
雪莉酒醋或紅酒醋	50公克	50 grams sherry vinegar or red wine vinegar

新派做法材料： / **Modern**

鮮奶油 UHT 35%	50公克	50 grams cream UHT 35% fat content
摩典納甜醋	50公克	50 grams modena vinegar
糖	100公克	100 grams sugar
水	50公克	50 grams water
錫箔紙		aluminium foil

Caramelized Stuffed Chicken with
Pinenuts and Sweet Modena

傳統 Traditional 什蔬松子燉雞肉做法：

1. 雞肉切丁，在砂鍋中煎至表面焦脆，取出。馬鈴薯和胡蘿蔔切丁（如同雞肉同樣大小）。

2. 將雞肉、馬鈴薯和胡蘿蔔放入平底鍋中一起炒至蔬菜表面焦黃，加入酒醋，煮至水分幾乎完全蒸發。

3. 將雞肉、馬鈴薯和胡蘿蔔放回鍋中，加清水至蓋住全部材料，煮到馬鈴薯變軟即可。上菜前，可加一些杏仁、松子和剁碎的荷蘭芹。

Tips：
若用紅酒醋，可以加少許糖來提味。
You can use a little more sugar
with red wine vinegar.

Cut the chicken into cubes. Fry it in a casserole until it's toasted and remove. Cube the potatoes and the carrot into the same size portions as the chicken. Transfer the chicken to a pan and fry with the vegetables until vegetables are toasted, then add the wine vinegar and reduce it until it's almost dry. Return the chicken to the casserole and cover with water. Cook everything together until the potatoes are tender. Before serving, add the almonds, the pine nuts, and some chopped parsley.

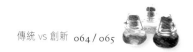

新派 Modern 松子雞肉卷做法：

1. 將削好皮、切成丁的馬鈴薯放入砂鍋中，加入清水蓋過馬鈴薯，煮到變軟為止。將水濾掉，馬鈴薯趁熱放進果汁機，加入鮮奶油和40公克的橄欖油打成泥。加入鹽和胡椒調味。

2. 胡蘿蔔削皮剁碎，也可以用果汁機打碎。加入10公克的橄欖油來做成"胡蘿蔔油"。

3. 將50公克的水和50公克的砂糖放進砂鍋中，用中火不停地攪拌煮成焦糖。焦糖變成咖啡色之後，加入摩典納甜醋煮5分鐘。取出，放入冰箱。

4. 把雞胸肉片開，如同一本打開的書，在攤開的肉上塞入松子、鹽和黑胡椒，將雞肉折好。拿一張錫箔紙，像包糖果一樣，兩邊捲起來，把雞肉包住。在一只平底鍋底放一些水，放入雞肉，用鍋蓋蓋住，慢煮約10分鐘。

5. 煮好將雞肉取出，將雞肉捲切成2塊，同時切去角。雞肉沾滿焦糖醋汁後，在一只非常燙的平底鍋中炙到表面焦脆。

6. 在盤子中間放一坨馬鈴薯泥，把雞肉放在上面，淋上一些摩典納焦糖醋汁和杏仁。再以胡蘿蔔油和荷蘭芹點綴。

Add the peeled, cubed potatoes to a casserole, cover with water and boil it until they are tender,Strain off the water. While the potatoes are still hot, put them in a blender, add the cream and 40 grams of olive oil, and blend it until creamy. Add some salt and pepper.

Peel and mince the carrot. You may use the blender for this step. Cover the carrots with 10 grams of olive oil to create a "carrot oil."

Put 50 grams of water and 50 grams of sugar in a small casserole and prepare a caramel by stirring constantly over medium heat. Once the caramel is brown in colour, add the modena vinegar and cook for 5 minutes. Remove and place the mixture in the refrigerator.

Fillet the chicken breast as if it were a book. Stuff the open breast with the pine nuts, some salt, and some black pepper. Fold the breast closed. Using the aluminium, prepare a wrapper and twist both sides closed in the manner of rolled candy. Cover the bottom of a fry pan with some water and place the chicken rolls inside. Cover and cook slowly for 10 minutes.

When done, remove from the pan and cut the chicken roll into two pieces. Also, cut the uneven corners off the roll. Dip the chicken into the sugar and vinegar mixture and--in very hot frying pan--caramelize the chicken.

Place a dollop of the potato cream in the centre of a plate and, over it, place the chicken. Drizzle the chicken with some sweet modena caramel and some almonds. Garnish with the carrot oil and parsley.

Stewed Meat, Mama's Style

有媽媽味道的燉肉

　　走到世界任何一個國家，最百吃不膩的味道，絕對是有親切記憶的媽媽味，台灣有紅燒肉，西班牙也有很傳統的紅酒香料燉肉。選材可以用豬或牛的頰骨肉，做法很簡單，所有材料處理好之後，用油香煎，再加紅酒、香料以慢火煨燉，很有「慢食 Slow cooking」的精神。

　　這個走到那裡都能召喚遊子思鄉情愁的食譜，我把它的造型和吃法摩登化，利用炸過的馬鈴薯做容器，分別裝上炒過的蔬果丁、燉肉，配上濃稠肉汁，端上桌就像一個美麗的玉製珠寶盒。

No matter where you go, you'll never get fed up with your mom's cooking or cuisines of Mom's flavor. Just as pork stew in brown sauce is for Taiwanese, so we have meat stew in red wine in Spain.

To cook meat stew in red wine you should use cheek meat, either beef or pork is fine. Properly handle all the ingredients. Sauté the meat with oil, add red wine and spices, and then slowly stew over low heat. This is a dish of slow cooking.

No matter where a traveler goes, this dish always evokes his or her nostalgia. I've tried to modernize the old recipe for this dish: use deep-fried potato as a container, put fried diced vegetables and stewed meat in, and serve with thick and rich gravy. It will look like a beautiful jewelry box.

Cooking time›› 30 minutes.

主材料：

牛頰骨	800公克	800 grams beef cheekbone
美國大馬鈴薯	2顆	4 big American potatoes
紅蘿蔔	200公克	200 grams carrot
新鮮青豆	200公克	200 grams fresh peas
番茄	1顆	1 tomato
青椒	1個	1 green pepper
紅椒	1個	1 red pepper
大蒜	2瓣	2 cloves of garlic
0´4° O & Co橄欖油		0´4° ANTONIO CANO brand olive oil
紅酒	1公升	1 litre red wine (your choice of table wine)
新鮮迷迭香	10公克	10 grams fresh rosemary
鹽、黑胡椒	各適量	salt & black pepper

基本料理：

1.將黑胡椒和鹽塗抹於牛頰骨上。在一只大的平底鍋中，用0'4°ANTONIO CANO 橄欖油將牛頰骨的表面煎至焦黃。將牛頰骨從鍋中取出，加入切好的蔬菜丁（除了馬鈴薯），將蔬菜慢慢煮熟。

2.取出所有蔬菜，將剩下的油留在鍋中，並將牛頰骨放回鍋中，加入紅酒（要小心可能會起火）。將紅酒煮至幾乎完全蒸發，之後加入清水直到蓋住所有的材料，再加入迷迭香，慢燉約 3小時直到肉變爛為止。

傳統 Traditional 媽媽味道的燉肉做法：

1.馬鈴薯切丁，用油炸至香脆金黃。將所有的材料放回鍋中，用大火滾10分鐘。

2.千萬不要用湯匙攪拌鍋中的料，以免 "骨肉分離" ；輕輕搖晃鍋子來混合所有的材料即可。

Put some salt and black pepper on the cheekbone. In a big pan, fry the cheekbone with the 0'4º ANTONIO CANO olive oil, until it's browned on the outside. Remove the cheekbone from the pan and add the diced vegetables (not the potatoes). Cook the vegetables slowly until tender.

Remove the vegetables, but leave the olive oil in the pan. Return the cheekbone and add the wine (be careful, as the mixture may flame). Reduce the wine until it's almost gone, then cover with water, add the rosemary and cook slowly until the meat is tender (around 3 hours).

Dice the potatoes and deep fry it until they are crispy and golden in colour. Return all the ingredients to the pan and boil for 10 more minutes.

Avoid using a spoon to mix the sauce and vegetables with the cheekbone because the meat will break apart. If you need to mix ingredients together, gently rock the pan.

新派 Modern 媽媽味道的燉肉做法：

1. 馬鈴薯削皮，切成 5 x 5 x 10 公分的大小。用熱油炸至變軟，再放入冰箱中冷卻。
2. 用去核器在馬鈴薯中間挖2個洞。在其中一個洞裡面放肉，另外一個洞裡面用漏斗將番茄、紅蘿蔔、青豆、青椒和紅椒依次放入。
3. 放入烤箱用攝氏200度烤 5分鐘直到表面焦脆。
4. 將湯汁淋上並用迷迭香葉裝飾即可。

Peel the potatoes. Cut into 5cm. x 5cm. x 10 cm rectangles. Deep fry them slowly until soft. Once fried, put them inside the refrigerator to chill.

Using an apple corer, make two holes inside the potato. In one of the holes, put the meat. Using a funnel, stuff the other hole with the tomato, then the carrot, then the peas, and—finally—the green and red pepper.

Put inside the oven for 5 minutes at 200ºc to toast the outside of the potato.

Serve the sauce over this and decorate with a rosemary leaf.

肉桂牛奶燉飯（米布丁）

　　西班牙東南部靠近地中海沿岸一帶盛產稻米，除了長米之外，也有跟台灣蓬來米類似的圓米，因為是魚米之鄉，當地有各種以稻米為主的名菜，瓦倫西亞（Valencia）的海鮮飯（Paella）就是最佳代表。

　　除此之外，米飯也普遍用於甜點當中，以牛奶、糖和米飯為主體的米布丁（肉桂牛奶燉飯）就是這裡的代表性甜點。米布丁的原始食譜出自教堂，早年居民們把家裡多餘的稻米和牛奶，送到教堂做為奉獻，修士們絞盡腦汁發展各式吃法，因而產生這道香甜柔滑的甜點。

　　傳統米布丁很好做，在新派吃法裡，我為它製造了一層酥脆的焦糖外衣，增加口感，再搭配一球冰淇淋，一冷一熱、有脆有軟，還有肉桂和檸檬清香，一點點嶄新的搭配靈感，翻新了米布丁傳奇。

Rice grows along the Mediterranean coast in south-eastern part of Spain. In addition to long grain rice, there grows short grain rice, too. Being the homeland of rice and fish, this region is famous for its rice cuisine; Valencia Paella is one of the best examples.

Moveover, rice is used for making desserts; rice pudding made from milk, sugar and rice is a typical dessert from this region. Recipes of rice pudding originated from the churches. In the early days people donated rice and milk to the churches in their parish which made monks rack their brains to come up with different ways to cook and eat rice. What came out was the sweet and smooth rice pudding.

While it's easy to make traditional rice pudding, I've tried to coat it with a layer of crispy caramelized crust to give it a new taste. Also, serve it with a scoop of ice cream. Thus, I combine hot and cold, crisp and soft, and cinnamon and lemon together. With a little inspiration, I've tried to reinvent rice pudding.

Cooking time›› 40 minutes.

主材料：

牛奶	1公升	1 litre milk
米	80公克	80 grams rice
肉桂棒	1支	1 cinnamon stick
糖	125公克	125 grams sugar
萊姆皮	少許	lime rind
肉桂粉	少許	cinnamon powder

新派做法材料： / **Modern**

牛乳冰淇淋	500公克	500 grams milk ice cream
米	80公克	80 grams rice
牛奶	200公克	200 grams milk
肉桂棒	1支	1 cinnamon stick
糖	100公克	100 grams sugar
水	20公克	20 grams water
萊姆皮	少許	lime rind
肉桂粉	少許	cinnamon powder

Rice with Milk and Cinnamon

傳統 Traditional 肉桂牛奶燉飯做法：

1. 牛奶倒入鍋中，加入肉桂棒和薄薄兩片（去掉白色部分的）萊姆皮，煮沸。

2. 牛奶一煮沸，倒入米，攪拌到沸騰。轉中火煮30分鐘，要不時攪拌。飯一軟就可以加糖，放冷後放進冰箱。

3. 用碗盛點牛奶燉飯，撒上一點肉桂粉，再撒點磨碎的萊姆皮。

Pour milk into a casserole, then add the cinnamon stick and the lemon rind (two slices with the white pulp trimmed away). Bring the milk to a boil, add the rice, and mix until it boils again. Next, cook over medium heat for 30 minutes, mixing occasionally. Once the rice is soft, add the sugar and transfer to the refrigerator. Serve in a bowl with some rice and milk. Dust with some cinnamon powder and top with a little grated lemon rind.

新派 Modern 肉桂牛奶燉飯做法：

1. 牛奶倒入鍋中，加入肉桂棒和萊姆皮。

2. 牛奶煮沸，倒入米，煮20分鐘，不時攪拌一下。

3. 將米糊倒在烘焙紙上，攤成薄薄一層（厚約1公分），放入冰箱冷凍，凍好即可取出，用模具切成圓形的米餅，撒上白糖。

4. 熱鍋，將米餅放入鍋中，兩面煎到焦糖化。

5. 用肉桂粉、糖、水，加一點萊姆皮，熬煮成焦糖。煮好，濾掉萊姆皮，變成褐色即可起鍋。

6. 用一只大匙反覆舀了又倒，倒了又舀，直到焦糖拔絲。用手拔絲，小心保持絲狀，用以裝飾。

7. 將已經焦糖化的圓形米餅擺到盤中，舀一匙冰淇淋放在米餅上，擱上焦糖絲，再撒上肉桂粉和研碎的萊姆皮。

Pour milk into a casserole, then add the cinnamon stick and the lemon rind. Once the milk boils, add the rice and cook for 20 minutes, mixing occasionaly. Spread the rice mixture on oven paper, creating a thin, 1cm layer. Transfer to the refrigerator. Once chilled, cut with a mould into round rice wafers. Sprinkle with white sugar and caramelize both sides in a very hot pan.

With the cinnamon powder and some lemon rind, prepare a caramel (see Saint's Bones for simple instructions). Once the caramel is ready, remove the lime rind with a sieve. Once browned, remove the caramel from the heat.

Using a tablespoon, repeatedly lift and pour the caramel over itself until--when drizzled--it retains a threadlike consistency. Next, roll the ropey caramel by hand, taking care to maintain the shape of the thread. Keep this caramel coil for decoration.

Place the round, caramelized rice cookie in the centre of a dish. Settle a spoon of the ice cream over the cookie and top with the caramel coil. Dust with cinnamon powder and garnish with grated lemon rind.

西班牙焦糖肉桂海綿蛋糕

「Torrija」是一道很經典的西班牙家常點心，也是消化吃不完麵包的好方法，它的做法很類似法式吐司，將麵包切片沾上牛奶、蛋液，再用油煎到外酥內軟，灑上糖粉和肉桂粉，做法雖簡單，但相信我，它的滋味好極了，很容易讓人一片接一片，停不了口。

在新派做法裡，我讓傳統的「Torrija」升級，改用海綿蛋糕取代麵包，讓口感更好，做出來的造型也更美觀，重點在浸飽牛奶的海綿蛋糕，沾上糖粒之後，再在熱鍋上煎到焦糖化，撒上肉桂粉，佐以濃濃的優格醬，即使嘴饞多吃幾個也不膩。

"Torrija"is a very classic home bakery in Spain, and also a good way to digest the unfinished bread with similar to French toast cooking method, slice the bread and dip with milk, egg cream and fry until it turns crispy outside and soft inside, spill with icing sugar and cinnamon powder, an easy way of cooking but with fabulous taste, believe me, it gonna make you eat one piece after another.

In new way of cooking, I have upgraded traditional "Torrija", with bread replaced with sponge cake instead to make a better taste and more pleasing to the eye, with a focus on sponge cake that fully sank with milk, and fry in a hot pot until it is caramelized after dipping with sugar, spill with cinnamon powder and serve with thick yogurt sauce, you will never get tired of eating with your greedy mouth.

Cooking time›› 30 minutes.

主材料：

法國長棍麵包	1條	1 French baguette
牛奶	500毫升	500 millilitres milk
糖	200公克	200 grams sugar
蛋	4個	4 eggs
肉桂粉	10公克	10 grams cinnamon
葵花油	100公克	100 grams sunflower oil

新派做法材料： **Modern**

葡萄乾核仁海綿蛋糕	200公克	200 grams raisin-nut sponge cake
牛奶	500毫升	500 millilitres milk
蛋黃	6個	6 egg yolks
糖	200公克	200 grams sugar
肉桂粉	10公克	10 grams cinnamon

Torrija

傳統 Traditional 西班牙焦糖 肉桂海綿蛋糕做法：

1. 麵包切成薄片，拌入牛奶裡面，靜置1個小時左右。
2. 打散蛋黃汁；濾去麵包上多餘的牛奶，再把浸溼的麵包加入蛋黃汁中。
3. 用葵花油煎香麵包，然後濾去多餘的油，最後，撒上糖粉和肉桂粉做裝飾。

Cut the bread into slices and mix with the milk. Allow the mixture to sit for around 1 hour. Meanwhile, mix the eggs. Next sieve off the excess milk from the bread and add the moist bread to the mixed eggs.

Deep fry this dough with sunflower oil. Sieve off the excess oil. To garnish, dust the bread with sugar and cinnamon.

新派 Modern 西班牙焦糖
肉桂海綿蛋糕做法：

1. 海綿蛋糕切塊。牛奶倒入鍋中，煮沸。
2. 砂糖拌入蛋黃中去打。再將（接近沸點的）熱牛奶倒入蛋汁中攪拌，拌到乳化。再加入一塊塊的海綿蛋糕，靜置1小時。
3. 待海綿蛋糕和牛奶混合之後，倒點糖在盤中，沾海綿蛋糕，把蛋糕都沾滿。
4. 熱鍋，把蛋糕煎到焦糖化，撒上肉桂粉，佐以英式奶油或濃濃的優格。

Cut the sponge cake into squares, then add the milk to a pan and heat until boiling. Meanwhile, mix the egg yolks with 100 grams of sugar.

Next, pour the hot milk (it should be close to boiling) over the egg mixture and stir until creamy. Now, add the sponge cake squares and allow it to sit for one hour.

After the cake and milk have had time to mingle, pour some sugar on a plate and dredge the sponge cake with it until covered; then, in a hot pan, caramelize it. Sprinkle some cinnamon on the top and serve with some of the English cream or rich yoghurt.

Saint's Bones, Updated

"改版" 聖人骨

　　西班牙是天主教國家,至今仍保有許多宗教節日,每年復活節前一週的「聖週」(Semana Santa),是西班牙最重要的宗教節慶之一,人們沿襲過去悠久的傳統,在這段時間舉行盛大的美麗花車表演,為了慶祝聖週,人們還會到商店買一種叫「聖人骨」(Saint Bones)的甜點,用杏仁、雞蛋和糖漿做的「聖人骨」,糖度很高,口感非常甜,有人形容那甜美的味道「好像進入天堂看到天使」一樣。

　　「聖人骨」也是一道從教堂中傳出來的甜點食譜,主材料杏仁和雞蛋是一般民眾經常送到教堂奉獻的供品,修士們利用它們研發變成極具特色的甜點,我把這道經典點心,結合上現代化的盤飾想法,重新在餐廳演繹推出,非常受到客人的歡迎。

Spain is a Catholic country and a lot of religious festivals are still kept until now. Semana Santa (Holy Week), the week immediately before Easter, is one of the most important rituals in Spain. During this week people will follow the old tradition and give a big flower parade. To celebrate the Holy Week, people will go to shops and buy Saint's Bones, a dessert made from almond, eggs and syrup. Saint's Bones are very sweet, so people describe the taste of the sweets "as if go to heaven and see the angels."

It's from the church that the recipe of Saint's Bones was passed down. The main ingredients, almonds and eggs, were the offerings people gave to churches and later monks turned them into a specific kind of sweets. This dish of classic dessert has been very popular with our guests after I recreated it with modern ideas of garnish.

Cooking time›› 1 hour.

主材料:

雞蛋冰淇淋:

蛋黃	10顆
水	100公克
砂糖	100公克

Egg ice cream:

10 egg yolks
100 grams water
100 grams sugar

杏仁醬:

杏仁	200公克
砂糖	100公克
水	100公克
烘焙紙	

Almond Paste:

200 grams almonds
100 grams sugar
100 grams water
oven paper

新派做法材料:

焦糖:

| 砂糖 | 50公克 |
| 水 | 25公克 |

Modern

Caramel:

50 grams sugar
25 grams water

傳統 Traditional
聖人骨做法：

蛋糕：

1.將水和砂糖放入平底鍋中用小火慢煮至變成糖漿但尚未變色。

2.將蛋黃放入果汁機打散，同時趁熱慢慢放入糖漿直到汁液完全混合均勻。放進冰箱備用。

杏仁膏：

1.再一次，將砂糖和水放入鍋中用小火慢慢煮到變成糖漿但未變色。

2.用果汁機將杏仁打成粉，慢慢加入糖漿直到變成膏狀。變冷前用兩張烘焙紙夾住壓平，用麵棍擀成0.25公分厚。

3.從杏仁膏切出長方形，塗上一層蛋糕，將杏仁膏捲起來，將做好的杏仁膏蛋捲放入冰箱。

Egg Paste

Cook the sugar and water together in a pan over low heat until the mixture becomes a syrup, but without changing colour. Then, while the syrup is still hot, mix the egg yolks in a blender, adding the syrup slowly until all is incorporated. Set aside in the refrigerator

Almond Paste

Once again, cook the sugar and water together in a pan over low heat until the mixture becomes a syrup, but without changing colour. Using a blender, mince the almonds until they become a powder, then add the hot syrup slowly until the mixture becomes a paste. Before the paste is cold, press it between two strips of oven paper and extend the dough using a rolling pin until 0.25 cm. in thickness. Cut rectangles from the pastry and fill with a line of egg paste, then roll the almond paste around the egg and store the rolls in the refrigerator.

新派 Modern "改版" 聖人骨做法：

1. 照傳統做法的方式做蛋糊，但做好之後放進冷凍庫。

2. 杏仁膏也依照傳統方式製作，做好之後將麵糰裝入烘焙袋放進冰箱。

3. 用水和砂糖熬焦糖，變成咖啡色之後關火。用湯匙不停將焦糖拉起攪拌，焦糖拉出一絲絲的感覺。接著用手將麻花狀的焦糖捲起來，要小心維持麻花的形狀。焦糖捲留著做裝飾。

4. 將杏仁糊放進平底鍋中，不停攪拌直到變乾脆。冷卻後放入果汁機打成粉。

5. 將螺旋花嘴置於杏仁膏中，用火槍炙燒表面。撒上杏仁粉，旁邊放一球冰淇淋，用焦糖捲做成裝飾。

To make the egg paste, prepare the modern version exactly the same way as the traditional, but once finished put the paste in the freezer.

For the almond paste, prepare the same way as the traditional method, but when finished, put the dough inside the refrigerator in a pastry bag.

Prepare a caramel by heating equal parts sugar and water in a pan and boiling until golden. Once browned, remove the caramel from the heat. Using a tablespoon, repeatedly lift and pour the caramel over itself until it becomes cooler and, when drizzled, it retains a threadlike consistency. Next, lift the caramel with a spoon and roll it around your hand like winding a thread, taking care not to break the elastic string. Keep this roll of caramel threads for decoration.

Place 1 tablespoon of the almond paste in a pan, stirring all the time until it´s toasted and dry. Once it´s cooled, whir in a blender to produce the almond powder.

Using the pastry tube, squeeze out a long line of pasty, doubling back on itself until you layer the pastry four folds high (in the shape of a cube). Next, brown the top of the cube with a blowtorch. Avoid tipping it over. Dust the roll with some of the toasted almond powder and serve with one small spoon of the ice cream. Decorate with the caramel coil.

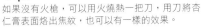

Tips：

如果沒有火槍，可以用火燒熱一把刀，用刀將杏仁膏表面烙出焦紋，也可以有一樣的效果。

If you don't have a blowtorch, heat a knife on the burner, then just touch the surface of the roll.

靈感乍現的創意菜
My Own Creations

從傳統出發，融合個人創作靈感的風格美食
Some of my creations, inspired on the Spanish food but with my personal touch…

椰子玫瑰西班牙櫻桃冷湯

Gazpachos 西班牙冷湯是一道最能表現西班牙熱情陽光的菜色。

西班牙南部的夏天，氣溫可以高達攝氏 40 度，從非洲吹上來的熱風，讓整個安達盧西亞像要燃燒起來一般，這麼高的溫度下，很難有好胃口，冰冰涼涼的冷湯，成為這個地區的居民安渡炎夏的最佳救贖。

傳統的安達盧西亞涼湯由番茄丁、黃瓜丁、青椒丁以及橄欖油、紅酒醋和大蒜製成，為了增加口感，通常會加入少量烤麵包丁。這道菜相傳有 600 多年歷史，據說是一位御廚為了討好當時統治安達盧西亞的國王，特別發明的菜式。我到台灣之後發現本地人普遍不習慣冷湯口感，其實台灣的夏季漫長炎熱，很適合享用冷湯。

多年前我在西班牙一間知名餐廳任職時，曾經為一位客人發明過一道「櫻桃冷湯」，那是一位超愛喝冷湯的老饕，要求餐廳一定要找到有機的紅番茄製作冷湯，正好那一陣子市場買不到好貨，為了滿足客人的要求，我召開廚房會議，大家集思廣益，想出用顏色、口感與番茄相近的櫻桃取代，基本做法及材料則都和傳統 Gazpachos 如出一轍，只是額外添加椰奶增加香氣和滑潤，再加有機的玫瑰花瓣做裝飾。

相信我，約會時為女朋友點這道菜，絕對能讓戀情加溫！

Gazpacho is a dish which best examplify the passion of Spanish sun.

In southern Spain the temperature may reach up to 40 degrees Celsius in summer. Hot wind blown upward from Africa makes Andalusia as if burning. Under such heat, people have no appetites. For local residents in this region a chilled cold soup becomes the best remedy for passing a scorching hot summer.

Traditional Andalusian gazpacho is cooked with diced tomatoes, cucumber, green pepper, garlic, olive oil and red wind vinegar, and usually some crouton is added to enhance the taste. It is said that some 600 years ago a royal chef invented this soup to please the king who then ruled Andalusia. After coming to Taiwan, I found people here are not used to eating cold soup. Actually, it's quite suitable to have cold soup for meals since summer is long and hot here.

Years ago I once created a cherry gazpacho for a guest when I was working at a famous restaurant in Spain. The guest, a glutton for gazpacho, asked the restaurant to make gazpacho with organic tomatoes. It happened that we couldn't find tomatoes of good quality then in the market. In order to meet the guest's request, I held a brainstorm meeting in the kitchen and came up with the idea of substituting cherries for tomatoes because of the similarities in color and taste of these two fruits. The ingredients and directions for making cherry gazpacho are basically the same as making tomato gazpacho, except with additional aroma and smoothness brought out by coconut cream and the garnish of organic rose buds.

Order this soup for your girlfriend when dating and, trust me, it will definitely warm up your relationship.

Cherry Gazpacho with Coconut
and Rose Blossoms

Cooking time›› 25 minutes.

主材料：

黑櫻桃	400公克	400 grams black cherry
青椒	1個	1 green pepper
紅椒	1個	1 red pepper
洋蔥	半個	½ onion
大蒜	3瓣	3 garlic cloves
黃瓜	20公克	20 grams cucumber
麵包	20公克	20 grams bread
水	50公克	50 grams water
白醋	25公克	25 grams white vinegar
0'4° ANTONIO CANO 橄欖油	100公克	100 grams olive oil (0´4º ANTONIO CANO brand)
鹽	10公克	10 grams salt
胡荽（香菜）籽	半茶匙	½ teaspoon coriander seed
玫瑰	2朵	2 roses
椰奶（藍象牌）	50cc	50 millilitres coconut milk (BLUE ELEPHANT brand)
蜂蜜	10公克	10 grams honey
荷蘭芹	50公克	50 grams parsley

椰子玫瑰西班牙櫻桃冷湯做法：

Cherry Gazpacho with Coconut
and Rose Blossoms

綠油：

將1瓣大蒜、荷蘭芹跟50公克的 ANTONIO CANO 橄欖油放進果汁機，打勻至所有材料變成汁，取出放入冰箱備用。

冷湯：

1. 把麵包放在一個碗裡，撒上一些水讓麵包軟化。將大部分的櫻桃去核，有的留下做裝飾。將去核的櫻桃、青椒、紅椒、洋蔥、大蒜、胡荽籽和黃瓜放進果汁機，用最高速打2分鐘。

2. 加入麵包，繼續用果汁機打勻，同時加入清水和醋，最後加鹽和剩下的橄欖油，再打2分鐘。將打好的冷湯濾渣，放入冰箱冷藏。

3. 盛盤前，先將一粒櫻桃放在馬汀尼杯裡，之後倒入綠油蓋住櫻桃。再在每個杯子裡倒入藍象牌椰奶，在杯緣擺上幾片玫瑰花瓣做裝飾。同時在另外一個馬克杯裡盛放冷湯，和綠油一起上桌。

First, prepare the green oil: place one garlic clove, the parsley, and 50 grams of ANTONIO CANO olive oil in a blender. Blend until the ingredients become an even cream. Reserve the mixture in the refrigerator.

Next, prepare the gazpacho: put the bread in a bowl and sprinkle with the water in order to soften. Remove the seeds from most of the cherries, retaining some for decoration. Add the pitted cherries the green peppers, red pepper, onion, garlic, coriander and cucumber to the blender and run at the maximum speed for two minutes. Next, add the bread, running the blender continuously while adding water, then vinegar, and, finally, the salt and the remaining olive oil. Blend it for two more minutes, then sieve the gazpacho and keep it in the refrigerator.

To serve, begin by placing one whole cherry at the bottom of a Martini glass. Cover the red fruit with the green oil and then slowly add some BLUE ELEPHANT coconut milk to each glass. Put some rose blossoms on the rim of each glass. Finally, serve together with the cold gazpacho in a mug. Serve.

椰奶蛤蜊濃湯
In Equilibrium

在餐廳的菜單上，我為這道菜取了一個有趣的名字「平衡桿」，比較有想像力的客人，一看到這道菜就會發出會心微笑，因為架在馬汀尼酒杯上那 3-4 顆蛤蜊，乍看之下正像是馬戲團裡，提著長竹桿走鋼索的空中高人呢！

這道菜造型給人走平衡木的聯想，它的味覺也力求一種平衡的和諧感，組成這道菜的三要素中——椰奶香甜、蛤蜊鮮鹹、南瓜奶油滑腴細膩，三種很不一樣的口感和香味，藉橄欖油巧妙融合在一起，提供口腔多層次的享受。

台灣的蛤蜊新鮮肥美，價格也不貴，這道菜的成本不高，只運用一些搭配和擺盤技巧，就能營造出讓人驚艷的感覺。學會了，下次宴客端上桌，包管大家對你的手藝另眼相看！

On the menu I called this dish In Equilibrium. Guests with imagination will smile immediately when they see the clams suspended over the martini. At first glance, it looks like a rope-walker with a long pole in his hands.

Not only does this dish give a visual association of balance beam, but it also strives for a balance of taste. The three elements of this dish—the sweetness of coconut cream, saltiness of clams and smoothness and richness of pumpkin cream—are blended perfectly together, through olive oil, and provide multiple sources of oral enjoyment.

Clams are fresh and pretty good here in Taiwan. Besides, they are not expensive. Frankly, with a matching side dish and garnish it will create an astonishing impression. Once you have learned to cook it, you can impress your guests with your culinary skills next time you give a feast.

Cooking time›› 25 minutes.

主材料：

義大利麵	4根	4 strands of dry spaghetti
蛤蜊	16個	16 clams
藍象牌椰奶	100公克	100 grams BLUE ELEPHANT brand coconut milk
南瓜	200公克	200 grams pumpkin
1°O & Co初榨橄欖油	50公克	50 grams virgin olive oil (1° O & CO brand)
洋香菜末	10公克	10 grams chopped parsley

做法：

1. 燒開水後放下義大利麵，約煮10分鐘到麵變軟。另外熱鍋，不加油，烘烤煮好的麵條，盛起，放乾燥處，不要收進冰箱。

2. 南瓜削皮，切塊，用橄欖油拌過，放入鍋中煮到熟軟。用果汁機將南瓜打成泥。

3. 再取一只鍋倒入椰奶和蛤蜊，蓋上鍋蓋，煮到蛤蜊開了，取出蛤蜊，去殼取肉（剩下椰奶蛤蜊高湯要留用），並用一根根烤得硬硬的義大利麵當竹籤去串蛤蜊肉，將蛤蜊肉推到麵條中段。

4. 舀幾匙南瓜濃湯，倒入馬汀尼酒杯中。將串著蛤蜊的義大利麵條擱在杯緣上，讓蛤蜊懸在濃湯上。撒上一點洋香菜，滴上幾滴橄欖油。趁熱上桌。配上一小碗椰汁蛤蜊湯，讓客人自行取用。

Cook the spaghetti in boiling water for 10 minutes until tender. Then in a hot pan, fry it without oil until toasted. Retain in a dry place (do not refrigerate).

In a casserole, toss the peeled, diced pumpkin in the O & CO olive oil and cook it until the pumpkin is tender. Transfer to a blender and purée to obtain a pumpkin cream.

In another casserole, stir together the BLUE ELEPHANT brand coconut milk and the clams. Cover and cook until the clams open. Remove the shells, sieve the juice and–using the stiff strands of spaghetti as brochettes-- skewer the clams and slide them to the centre.

Place a few spoonfuls of the pumpkin cream in the bottom of a Martini glass. Suspend the clams over the purée by resting the two ends of the spaghetti skewer on the rim of the glass. Decorate with some parsley and some drops of O&CO olive oil. Serve as soon as possible. Accompany with a small bowl of the coconut soup and encourage guests to help themselves.

Warm Clams and Purple Potato Salad
with Mint and Figs Vinagrette

熱蛤蜊紫薯沙拉佐薄荷無花果醋

　　這是一道做法簡單又充滿大自然氣息的菜餚，美味的關鍵在於：好的摩典納陳年酒醋和頂級初榨橄欖油，它們能為沙拉帶來不凡風味，酒醋部分我特別挑選了無花果香味的甜醋，橄欖油則用了帶薄荷香味的，它們為這道前菜帶來相乘的口感效果。

　　這道菜還有一項秘密武器，是利用煮蛤蜊剩下的高湯提鮮，在餐廳比較專業的作法，我會用發泡器把蛤蜊高湯打成柔細的泡泡，陪襯在沙拉旁，吃起來若有似無的鮮味迴盪在口腔，非常奇妙。食譜裡我提供比較簡單的做法，直接就把蛤蜊高湯加進橄欖油醋汁裡，也可以得到相似的口感。

This is recipe with simple way of cooking and full of atmosphere of nature, with key to a pleasant taste lies in : good Modena wine vinegar and top extract virgin olive oil will enable a different flavor to salad, while I particularly select sweet vinegar with fig fragrance, olive oil with mint flavor, to enable an equal mouthfeel to this appetizer.

This course has another secret weapon - use clam soup stock left to enable a fresh taste, a professional way of doing this in the restaurant is to use mixer to whip the clam soup stock into fine bubbles and serve as a contrast beside the salad, a taste swings in the mouth cavity, a feeling of wonder. I provide a more simple way of cooking in the recipe, where I directly add the clam soup stock into olive oil ginger and enable a similar mouthfeel.

Cooking time›› 20 Min.

主材料：

蛤蜊（大的）	16個
紫薯	4個
嫩綠雜菜沙拉	
O & CO 牌無花果味摩典納甜醋	10公克
O & CO 牌薄荷橄欖油	40公克

16 large clams
4 purple potatoes
Baby greens mixed salad
10 millilitres balsamic modena condiment
　fig flavoured oil (O & CO brand)
40 millilitres olive oil with mint (O & CO brand)

做法：

1. 紫薯用保鮮膜層層包住，用微波爐煮至變軟。再打開保鮮膜放冷，變成室溫的時候去皮。
2. 蛤蜊放入鍋中，加水蓋上鍋蓋煮至殼打開。煮熟之後將蛤蜊放入冰箱，不要將蛤蜊肉和殼分開。煮蛤蜊的水留著備用。
3. 用橄欖油、摩典納甜醋和鹽（約2公克就應該夠了，因為蛤蜊本身有鹹味）做成油醋。加入煮蛤蜊的水，混合至勻。
4. 用圓的模型切割去皮的紫薯，將切好的紫薯放在盤上。用竹籤將蛤蜊肉串起來，3~4個一串，之後將串好的蛤蜊插在紫薯上。放一些嫩菜沙拉配在旁邊，淋油醋在紫薯和沙拉上調味。

Wrap the purple potatoes in several layers of plastic wrap. Cook in the microwave until tender. Open the plastic carefully and let stand. Peel the potatoes when they have returned to room temperature.
In a covered fry pan, steam open the clams with a little bit of water. Once the clams are open, put them in the refrigerator, leaving them on the shell. Save the clam juice from the pan for the vinaigrette.
Prepare a vinaigrette with the olive oil and the modena vinegar, adding some salt (2 grams should be enough, because the clams are already salty). Add the reserved clam juice. Mix everything until the juice is incorporated.
Cut the peeled purple potato with a round mould and place the morsel on a plate. Serve the clams on a skewer, assembling three or four together as a brochette. Stick the skewer in the centre of the potato. Arrange some of the baby greens beside it and dress salad and skewer with the vinaigrette.

Pate De Foie-Gras with Greek Yogurt,
Orange and Green Tea Caviar

嫩鵝肝佐希臘優格&
橘子綠茶魚子醬

　　這是一道我為餐廳「EL TORO」特別研發的開胃菜，其中最有趣的地方是用綠茶仿製的綠色魚卵，這又是一個分子廚藝開創出來的美食魔術，利用高科技的原料和酸鹼平衡的原理，Ferran Adria 在他開設的 El Bulli 餐廳，創作出外形幾可亂真的鮭魚卵，客人小心用湯匙將魚卵送進嘴巴，牙齒輕輕咬破表面薄膜，噴在舌尖的卻是新鮮橘子、草莓或番茄汁，無數被這招騙過的客人莫不嘖嘖稱奇。

　　在「EL TORO」，我把鮮果汁換成綠茶，利用茶葉的清香平衡鵝肝的豐腴。在餐廳製作這道菜時，我們都是自己在廚房準備鵝肝和橘子醬，但這樣做非常花時間，而且我們使用的材料和工具價錢都很貴，所以我建議有興趣在家試做的人，直接利用吉利丁和水、油分子比重不同，也可以玩這手「幾可亂真」的模擬遊戲，唯一要注意的是使用的油顏色和口味必須清淡。

　　此外，橘子果醬和鵝肝醬都可以用買的，不過記住，一定要買最頂級的材料，做出來的口感效果才會好。

　　This is a appetizer that I develop especially for the restaurant, in which the most interesting part here is to use green tea to imitate the green roe, which is another magic created from molecular gastronomy, which utilizes high-tech ingredients and acid - alkaline balance principle, In El Bulli restaurant opened by Ferran Adria, where he created salmon roe that is just similar to real one, customers carefully put the roe into the mouth with spoon and gently bite through the thin membrane with teeth, what emits from to the tip of the tongue is fresh tangerine, strawberry or tomato juice, in which numerous customers that being fooled are quite astonished by this trick.

　　In "El Toro", I change the fresh juice into green tea and use the delicate fragrance of tea leave to balance the strong goose liver sauce. As we all prepare the goose live sauce and tangerine sauce alone in the kitchen while cooking this course in the restaurant, a very time-consuming process actually, also, the materials and tools that we use are very costly, therefore I recommend those trials at home would directly use different molecular of Gelatin, water and oil, they may play with simulation game of "similar-to-real" with hands, the only thing that they need to attend to is oil color and a light taste.

　　Besides, tangerine sauce and goose liver sauce could be purchased from outside, but you need to remember, you need best ingredients to deliver a good mouthfeel.

Cooking time›› 2 hours.

主材料：

頂級鵝肝	150公克	150 grams pâté de foie-gras (high quality)
希臘優格（全脂、濃稠）	80公克	80 grams Greek yogurt (whole, creamy yoghurt)
橘子果醬	60公克	60 grams orange marmalade
綠茶粉	10公克	10 grams green tea powder
綠茶	1包	1 bag of green tea
明膠	8公克	8 grams of gelatine
鹽	2公克	2 grams salt
水	100公克	100 grams water
任何清淡的種籽油 （如葵花籽油，勿用芝麻油）	100公克	100 grams of seed oil—i.e. 　　sunflower--(note: do not use sesame oil)
迷迭香葉	少許	rosemary leaves

嫩鵝肝佐希臘優格
&橘子綠茶魚子醬做法：

Pate De Foie-Gras with Greek Yogurt,
Orange and Green Tea Caviar

1. 明膠放入水中加些冰塊使它溶化。用100公克的熱水泡開綠茶包，趁熱加入綠茶粉、一點鹽和明膠，混合直到所有材料化開。用一支針筒吸滿綠茶，一滴一滴慢慢滴到油裡。綠茶碰到油的時候，會形成一粒圓珠狀。大約準備20粒圓珠，過濾之後再重頭來過，將做好的圓珠放在一個碗裡，用油包住。

2. 挖15公克（約1咖啡匙）橘子醬放進一個馬汀尼杯裡。鵝肝切成約1公分大小的方塊，放在橘子醬上面。用優格蓋住大部分的鵝肝，露出一塊。

3. 綠茶魚子醬將油濾掉之後用水清洗。將魚子醬放在優格鵝肝上，用迷迭香葉做裝飾。吃的時候，鵝肝、果醬、優格和魚子醬會融合成一股豐富、濃密、多層次的口感。

Dissolve the gelatin in cold water with some ice cubes. To make the caviar, prepare a cup green tea using the tea bag and 100 grams of water. Once it´s warm, add the green tea powder and a little bit of salt and the gelatin mixture. Mix everything until ingredients dissolve. Then, using a syringe draw up a cylinder of the tea mixture and, drop by drop, add it to the oil. Once a drop hits the oil, it will become a sphere as the gelatin interacts with the tea mixture, giving the pearl a stable structure. Prepare around 20, and then pass through a sieve and start again, placing completed spheres in a small bowl covered with oil.

In a martini glass, spoon 15 grams of the orange marmalade (about a coffee spoon). Cut some square pieces (approximately 1 cm. cubed) of the foie-gras and layer over the orange marmalade. Cover most of the foie-gras with the yoghurt, but leave one piece showing.

Sieve the caviar from the oil and wash with cold water. Top the yogurt and foie-gras with the caviar pearls and garnish with rosemary. When eaten together, foie-gras, marmalade, yogurt and caviar blend to create an incredibly rich and creamy harmony of flavours.

Tips：

在餐廳我們都自製橘子果醬和鵝肝醬，但這樣做很花時間，而且我們使用的工具和材料都很貴，所以我建議一般人可以直接用買的，不過切記：一定要買最頂級的材料！才能確保品質。

At my restaurant, EL TORO, we prepare the homemade pâté de foie-gras and the orange marmalade in our own kitchen, but it is time consuming and the equipment we use is expensive, so my suggestion is to buy these ingredients. Always ask and for the best quality.

Sealed Goose Liver with
Blub Lemon Cake and Black Honey

嫩煎鵝肝&
藍色檸檬蛋糕佐黑色蜂蜜

　　這是我到台灣之後創作的一道新菜，最詭絕的地方在於那塊艷藍色的檸檬蛋糕，美食的世界裡幾乎找不到這種藍色，當我在雞尾酒的國度裡，發現藍柑橘酒的時候，心裡不禁大聲歡呼：Bingo！美麗的藍柑橘酒一方面可以為蛋糕增添柑橘清香，另一方面也為這道菜多添了綺麗色彩。

　　藍柑橘酒是葡萄牙外海一個小島上的特產利口酒，當地盛產柑橘，所做的柑橘利口酒特別出名。我把海綿蛋糕整塊浸透在利口酒和萊姆汁當中，讓蛋糕吸飽汁液，再將煎好的鵝肝移到藍蛋糕上，讓鵝肝中的油脂滴下，交融滲透到蛋糕裡，最後放上藍象牌的檸檬香茅，淋上濃縮肉汁，激發出檸檬香茅的清香。

　　一道菜裡有美麗而神秘的顏色，有多層次的柑橘、萊姆及檸檬香茅清香，還有軟濡香滑的多元口感，像一份層層打開充滿驚喜的禮物。

I had created this new dish before I came to Taiwan. The most bizarre part of this dish lies in the blueness of the lemon cake since you can hardly find this kind of blue color in the culinary world. "Bingo!" I acclaimed loudly when I first observed blue curacao in cocktails. Not only does blue curacao add a gorgeous color to the cake, but it also brings a fresh smell of orange.

Blue curacao is a specific kind of liqueur from a small island off Portugal. Oranges grow abundantly on the island and liqueur made from those oranges is especially famous. I soak sponge cake thoroughly in the liqueur and lime juice, allowing the cake to fully absorb the juice. Then I put the sautéed goose liver on top of the blue cake, allowing the grease to drip down into the cake. Finally, I top the cake with lemon grass and drizzle the condensed gravy over, releasing the fresh smell of lemon grass. With its mysteriously beautiful color, multi-layered smell (from orange, lime and lemon grass), and soft and smooth taste, this dish is like a surprising gift with layers of layers of wrap to open.

Cooking time›› 30 minutes.

主材料：

新鮮鵝肝	400公克	400 grams fresh goose liver
海綿蛋糕	100公克	100 grams sponge cake
（義大利水果蛋糕或葡萄乾／果仁海綿蛋糕）		(panettone or a raisins/nut sponge)
藍柑橘酒	100公克	100 grams curaçao blue liquor
萊姆汁	10公克	10 grams lime juice
新鮮墨魚汁	20公克	20 grams fresh squid ink
蜂蜜	40公克	40 grams honey
鹽	10公克	10 grams salt
黑胡椒	10公克	10 grams black pepper
砂糖	10公克	10 grams sugar
乾藍象檸檬草	1片	1 leaf dry (Blue Elephant brand) lemongrass
濃縮肉汁	40公克	40 grams meat stock

嫩煎鵝肝
&藍色檸檬蛋糕佐黑色蜂蜜做法：

Sealed Goose Liver with Blub Lemon
Cake and Black Honey

1. 將新鮮鵝肝切成4塊，置於冰箱外直到鵝肝變成室溫。

2. 將萊姆汁和藍柑橘酒混合，把海綿蛋糕切成4塊之後，放入混
   合好的藍汁中浸泡至少20分鐘。其間要將蛋糕翻面讓蛋糕能
   夠均勻地吸滿汁液。

3. 用微波爐將墨魚汁加熱，之後放入果汁機，加入蜂蜜，用高
   速打勻直到完全變成黑色，用濾網過濾之後放進冰箱冷藏10
   分鐘。

4. 將鹽、砂糖、黑胡椒拍抹於鵝肝上，再用一只很熱的平底鍋
   將鵝肝煎熱。

5. 用一個乾淨的果汁機或食物調理機，放入藍象檸檬草和海
   鹽，打勻直至檸檬草變成粉狀。

6. 將藍色檸檬蛋糕置於盤子中間，並將煎好的鵝肝放在蛋糕
   上，用湯匙淋上黑色蜂蜜，並撒上打好的檸檬草鹽粉。

7. 用一片藍象檸檬草做裝飾，再淋上熱燙的濃縮肉汁即可。

Cut the goose liver into four pieces and leave it un-refrigerated until it returns to room temperature. Meanwhile, mix the lime juice and the curaçao. Cut four square pieces of the sponge cake and put them inside the blue sauce obtained by combining the curaçao and lime. Allow the cake to sit for 20 minutes at least, turning sometimes in order to moisten all sides.

Warm the ink in the microwave and put it in a blender. Add the honey, blending it at maximum speed until it´s completely black. Strain the paste and keep it in the refrigerator for 10 minutes.
Pat the salt, sugar and black pepper on all sides of the goose liver and seal it in a very hot pan.
In a dry blender, add the BLUE ELEPHANT lemongrass and some sea salt, blending it until the lemongrass becomes a powder.

Place the blue lemon cake in the centre of a dish, settle the goose liver over it, and then spoon the black honey over the liver and dust with some lemongrass salt powder.
Garnish with one leaf of BLUE ELEPHANT lemongrass and serve the meat sauce over it.

Sealed Scallop with Tomato Powder,
"Gallega" Garnish and Mandarin Oil

加里亞式炙干貝佐番茄粉

　　炙干貝這道菜在餐廳很受歡迎，它的烹調方式非常自然簡單，新鮮的北海道大干貝以乾鍋快速炙過，炙到兩面金黃就可以起鍋，由於干貝本身就帶有海水的鹹味，連鹽都不需添加，我用紅色番茄粉佐搭它的鮮美，另外配上用柑橘油炒香的洋蔥、大蒜和胡蘿蔔碎，吃來清新又爽口。

　　這也是一道在家很容易上手的料理，在餐廳我用進口紫色馬鈴薯做為配菜，一般人不容易買到紫色馬鈴薯，所以食譜上改用吐司麵包，配起來一樣秀色可餐。

　　Seared scallop is very popular at our restaurant. It's easy to cook; sear large fresh Hokkaido scallops in a pan without oil and remove them from heat until golden. Since scallop is naturally salty itself, salt is not needed. I use tomato powder to accentuate its freshness. Serving with fried minced onion, garlic and carrot, the scallops taste refreshing.

　　This is an easy-to-cook dish, too. At my restaurant for side dish I use purple potatoes which are not easy to get; therefore, I substitute toasted bread for purple potatoes. They look appetizing all the same.

Cooking time›› 10 minutes.

主材料：

北海道大干貝	4個	4 large Hokkaido scallops
紫洋蔥	1個	1 purple onion
胡蘿蔔	1條	1 carrot
大蒜	1瓣	1 garlic clove
月桂葉	1片	1 bay leaf
O & Co番茄粉	20公克	20 grams dry tomato powder (O & CO brand)
吐司（4片）	40公克	40 grams toasted bread (4 slices)
菠菜	100公克	100 grams spinach
Antonio Cano 0´4° O & Co橄欖油	40公克	40 grams olive oil
O & Co柑橘油	50公克	(0´4° ANTONIO CANO & CO brand)
橘皮		50 grams Mandarin oil (O & CO brand)
		orange rind

做法：

1. 干貝洗淨（我喜歡保留內臟，不過這是個人偏好），用溼紙巾包起來。
2. 加里亞式配菜：洋蔥、大蒜和胡蘿蔔切碎，用柑橘油去炒，炒到洋蔥變金黃色。
3. 橄欖油倒入鍋中，油熱了放下菠菜快炒，要炒得脆脆的。
4. 另取一只不鏽鋼鍋，加熱到很燙，不加油乾炙干貝，兩面都煎。由於干貝本身就有鹹味，所以不需加鹽。
5. 放一片吐司在盤中，抹上一大匙加里亞配菜，再擺上干貝。用幾片橘子皮做裝飾。並用番茄粉在干貝吐司旁邊畫出線條，上面再綴以菠菜。

Clean the scallops properly. I prefer to keep the roe or "coral," but this is a matter of personal preference. After cleaning the shellfish, wrap it in a damp napkin.

Gallega garnish: Mince the onion, the garlic and the carrot and cook slowly with the O & CO mandarin oil until the onion is gold in colour.

In a small casserole, add the olive oil and, once the oil is hot, fry the spinach until crispy.

In a hot pan without oil, seal both sides of the scallops. You don´t need to add salt, as the scallop is naturally salty.

In the centre of a plate, serve one slice of the toasted bread. Spread one tablespoon of the "gallega" garnish on the toast and top with the sealed scallop. Dress with some slices of orange rind. Draw a line beside the toast and scallop using the O & CO dry tomato powder and decorate with the crispy spinach on the top.

百香美乃滋藍鸚嘴魚佐魚鱗酥片

一般人吃魚不吃魚鱗，總是被刨進垃圾桶裡的魚鱗為什麼不能拿來吃？就是這樣一個念頭，促使我創作了這道菜。新鮮的藍鸚嘴魚用熱鍋香煎，魚鱗則留下來酥炸，炸得薄而脆，配著吃，好吃極了！

Normally, people eat the fish without the scales. Why can't we eat the scales that are always wrapped into the garbage can? It's this idea that makes me to create this dish. Fresh blue parrotfish is sautéed in the hot pan while the scales are remained to be deep-fried with thinness and crispiness. They are marvellous!

Cooking time›› 20 minutes.

主材料：

藍鸚嘴魚(保留魚鱗)	800公克	800 grams blue parrotfish (with scales on)
綜合蔬菜油	100公克	100 grams of some variety of seed oil
鹽	2公克	2 grams salt
黑胡椒	2公克	2 grams black pepper

百香美乃滋：

		Passion fruit mayonnaise:
百香果果醬	40公克	40 grams passion fruit jam
橄欖油	50公克	50 grams olive oil

做法：

1. 將魚清洗乾淨，去除魚鱗後並以乾紙巾覆蓋。將魚身切四等塊。
2. 製作美乃滋：先將百香果果醬倒入一只碗中，一邊加入橄欖油一邊使用手動攪拌器將果醬攪拌均勻。待橄欖油完全拌勻後，放入冰箱冷卻。
3. 製作魚鱗酥片：高溫熱油鍋直到油煙冒出，再加入魚鱗片油炸10秒鐘。將魚鱗酥片離鍋後置於一旁，以紙巾輕輕拭去多餘油脂。
4. 將魚煎至半熟，先煎魚皮表面，再煎另一面。離鍋後加入鹽及胡椒調味，再放入烤箱以攝氏180度烤5分鐘直到全熟。
5. 將魚放在盤中，並將魚鱗酥片放在魚肉上搭配百香果美乃滋食用。

Clean the fish and remove the scales, but keep it covered with a dry napkin. Cut the fish into four equal pieces.

For the mayonnaise, begin by placing the passion fruit jam in a bowl. Using a hand blender, mix the jam while adding the olive oil slowly. Blend until the oil is completely incorporated, then chill.

For the eatable scales, add the seed oil to a pan and heat until is almost burned. You will know it is hot enough when the oil begins to smoke. Now add the scale and fry for 10 seconds. Remove the crust from the pan and pat the oil away with a paper napkin.

Cook the fish to the point of being about halfway done in a pan. First, fry skin-side down, then fry the other side. Remove the fillets from the pan add the salt and pepper and transfer to an oven and bake slowly at 180º for 5 minutes more until is done.

To serve, place the fish on a plate. Top the fish with the scales and accompany with passion fruit mayonnaise.

Blue Parrotfish with Eatable Scales
and Passion Fruit Mayonnaise

Sauted Strawberry with Pink
Pepper and Honey Cookie

紅胡椒煎草莓配蜂蜜餅乾

一般人吃水果很少跟辛香料做結合，但是我發現紅胡椒跟很多水果都非常 match，我曾經用紅胡椒搭襯巧克力和鳳梨，那是一種非常濃郁熱情的南洋味 道；但是一樣的紅胡椒，配上草莓和蜂蜜餅乾，就有了不一樣的細緻風味，少 少的辛辣和紅胡椒鮮紅的色澤，為這道甜點單純的溫柔增加了戲劇性張力！

Rarely do people eat fruits with spices, but I find pink pepper go with a lot of fruits well. I used to pair chocolate and pineapple with pink pepper and it turned out to be rich in flavors of Southeast Asia. However, pink pepper with strawberries and honey cookies provides a delicate flavor. The spiciness and redness of pink pepper creates dramatic tension in this simple dessert.

Cooking time›› 30 minutes.

主材料：

草莓	12粒	12 strawberries
優格	100公克	100 grams yoghurt
紅胡椒	10公克	10 grams pink pepper
蜂蜜	100公克	100 grams honey
奶油	125公克	125 grams butter
蛋白	100公克	100 grams egg whites
中筋麵粉	100公克	100 grams all purpose flour

做法：

1. 草莓洗淨，切成一塊塊備用。
2. 融好100公克奶油，倒進攪拌機裡，加入麵粉，打勻，然後再倒入蜂蜜，最後再加入蛋白。
3. 將麵糊攤到烘焙紙上，把麵糊放進預熱至攝氏180度的烤箱，烤到金黃（大約5分鐘左右）。從烤箱取出，趁熱將它切成個人喜愛的形狀（例如圓形、瓦片狀、半球形、煙斗狀等）。把餅乾收到乾燥處。
4. 熱鍋，倒入25公克奶油融化，用大火嫩煎草莓，煎30秒即可。然後加入紅胡椒，起鍋前放入優格。馬上將草莓優格淋在脆脆的蜂蜜餅乾上，端上桌吃，最好吃。

Wash the strawberries and cut into quarters. Melt100 grams of butter and transfer to a blender. Add the flour and mix until completely incorporated, then add the honey and, finally, the egg white.

Spread the batter on oven paper and bake at 180ºc until golden (approximately 5 minutes.). Remove from the oven and, while still hot, cut the pastry into your preferred shapes (i.e. circles, tiles, domes, pipes). Store the cookies in a dry place.

Add 25 grams of butter to a hot pan, melt and sauté the strawberries on high heat for 30 seconds. Then add the pink pepper and set aside while the peppers warm. Next, add the yogurt.

Serve immediately on the crispy honey cookie. For best results, serve at the table.

烤鳳梨與辛香巧克力佐火燒蘭姆酒
Toasted Pineapple with Spicy Chocolate Cookie and Rum Flambe

　　鳳梨、巧克力、紅胡椒粒、蘭姆酒，看起來不相關的幾樣食材，放在一個甜點盤裡，轟！迸出不一樣的火花！這是我認為烹飪最迷人的地方——充滿各種可能性，像一個有趣的化學實驗。

　　在這道甜點中，鳳梨的微酸和紅胡椒的微辛，正好平衡巧克力的甜，各種口感在味蕾上達到一種巧妙的平衡。最炫的是，它是一道提供桌邊服務的甜點，美麗的烤鳳梨送上桌之後，由服務生淋上蘭姆酒，畫根火柴，調暗室內光線，鳳梨燃起熊熊火光，等酒精燒盡，空間裡飄送著烤鳳梨和蘭姆酒的香氣，是一道視覺、嗅覺、味覺兼具的幸福甜點。

Pineapple, chocolate, pink pepper and rum—those ingredients look irrelevant. But, put them on a dessert plate. Bang! They create sparks! Well, I think that's the fascination of cooking—full of possibilities, like an interesting experiment.

The slight sourness of pineapple and the slight spiciness of pink pepper just balance the sweetness of chocolate; different flavors reach a perfect balance on the taste buds. This dessert provides a chance for Gueridon service—the waiter must pour the rum, light a match, dim the light at the table after the pineapple is being presented. Then the pineapple will be burned in flames and send out mixed aroma of toasted pineapple and rum after the alcohol is burned out. This is a dish of vision, smell, and taste combined together.

Cooking time›› 20 minutes.

主材料：

鳳梨（大的）	1顆	1 large pineapple
黑巧克力（可可成份70%）	200公克	200 grams dark chocolate (70% cacao)
紅胡椒	20公克	20 grams pink pepper
帶鹽奶油	50公克	50 grams butter (salted)
直徑3公分的奶油餅乾	4片	4 cookies (3 centimetres in diameter)
砂糖	50公克	(butter cookies are the best)
水	50公克	50 grams sugar
黑色蘭姆酒	100cc	50 grams water
		100 millilitres black rum

做法：

1. 用圓形模型將鳳梨切成直徑3公分、5公分厚的形狀，用去核器在鳳梨中間挖一個洞。將剩下的鳳梨打成汁，過濾，加入砂糖和水煮成糖漿，保溫1小時。

2. 用中火將紅胡椒跟奶油加熱2分鐘，之後準備另一只鍋子放入熱水，將煮奶油胡椒的鍋子放進水裡隔水煮。加入巧克力，煮至巧克力溶化，保溫。

3. 在一只熱鍋中，翻煮鳳梨直到變成金黃色。將熱的鳳梨放在一個湯盤的中間。接著用漏斗將巧克力醬放在鳳梨中間，再將一片跟鳳梨同樣大小的餅乾放在鳳梨上。在一只鍋中燒熱一點蘭姆酒（做的時候要小心），淋在鳳梨上。在上菜前用火點燃蘭姆酒，火焰熄滅後即可享用，吃的時候小心燙！

Cut the pineapple using a round mould 3 cm. in diameter and 5 cm. thick. Make a hole in the centre of the pineapple using an apple corer. Juice the rest of the pineapple, sieve, and combine with the sugar and water. Cook the mixture down to a syrup, then add the pineapple to the syrup and keep it warm for one hour.

Heat the butter with the pink pepper for two minutes over medium heat. Next, prepare a bain marie (double boiler) with warm water. Place the pan containing the melted butter and pepper in the heated water. Add the chocolate and mix slowly until the chocolate is melted. Keep it warm.

In a hot pan, stir the pineapple until it´s golden. Serve the pineapple hot in the centre of a soup dish. Then, using a funnel, put the chocolate inside the hole in the pineapple, top with a cookie with a diameter equal to the pineapple round. Slightly heat a little bit the rum in a pan and--being extremely careful–serve over the pineapple. When ready to serve, flame the rum. After the alcohol burns off, it's ready to eat, hot!

融合的美味
Fusion

在這個美食的新領域中，我和我的二廚張
榮恩聯手，結合兩種文化創造出美食新感
受，感謝帶領我走進台灣奧秘的朋友！

The next frontier, to mix both cultures and obtain
something new, with the invaluable collaboration of my
sous chef, John, thanks my friend for showing me the
secrets of Taiwan

西式茶葉蛋

　　我從西班牙來到台灣之後，很快融入這裡的生活，也發現許多有趣的在地小吃，每個便利商店都可以買到的茶葉蛋就是其中一項。

　　茶葉蛋是充滿台灣味的記憶，茶葉和醬油交融的香氣，為水煮蛋添上了彩紋和滋味，常買茶葉蛋來吃的我，有一天突發奇想：為什麼不利用這麼在地的口味創作一道新菜？因而產生這道具有融合趣味的「西式茶葉蛋」。

　　分子廚藝這幾年在西班牙很夯，2007年的時候西班牙廚師又發展出一道名為「流金」（Liquid Gold）的創意菜，這道菜端上桌的時候看似一枚金蛋，但用刀輕輕一劃，蛋黃卻像艷陽下融化的奶油一般流淌下來，充滿華美的感覺。這是一項新的烹飪技術，主要是讓一枚雞蛋的蛋白凝固，蛋黃卻完全保持在液體狀態下。我把這項新烹技和茶葉蛋做結合，奇異的滋味讓很多人吃了大呼不可思議。

　　其實一般人也可以在家玩玩這種東西合璧的趣味，不必去學高深的新烹技，只要善於掌握蛋白在攝氏65度的水溫中將凝未凝的效果，就可以做出相似度90分以上的溫泉蛋。至於茶葉蛋中的紅茶、醬油、八角和滷包，我把它們分別變成醬油泡泡、紅茶吉利丁和香料餅乾，當你劃開蛋白，倒進配料並且一口送進嘴巴的時候，一種似曾相似的茶葉蛋記憶，準確出現在味蕾之上……。

Quickly, I have blended into local life in Taiwan after arriving from Spain. Also, I found many interesting local food here and tea eggs sold at convenience stores everywhere was one of them.

Tea egg with mixed aroma of tea and soy sauce is full of Taiwanese flavors and colors. One day an idea occurred to me, who frequently bought and ate tea eggs. Why don't I use such a local flavor to recreate a new dish? Thus a fusion dish of Western-style Taiwanese tea egg was created.

Molecular gastronomy has been quite popular in Spain these years. In 2007 a Spanish chef presented a so-called Liquid Gold creative dish. When the seemingly golden egg is slashed open after serving, the egg yolk drip down like butter melted under the bright sun. It's a new culinary skill to keep the egg yolk in liquid state while the egg white still solid. I've used this skill to make my western-style tea egg and the taste incredibly surprise people who eat it.

Actually, you can cook fusion dishes at home just for fun, without learning sophisticated culinary skill. If you can cook the egg in 65 degree Celsius warm water and thus keep the egg white half-fluid, you'll be a master of pouch egg. As for the black tea, soy sauce, star anise and marinade, I've turned them into tea jelly, soy bubbles and herbal cookies. When you slash open the egg white, mix the ingredients together, and put them into your mouth, your taste buds will experience a sense of deja-vu...

Black Tea Egg the Western Way

Cooking time›› 2 hours.

材料：

蛋	4個
紅茶包	2袋
醬油	150公克
小條法國長棍麵包	1條
八角（大茴香）	1粒
小茴香籽	5公克
柳橙	1個
吉利丁片（16公克）	2片

4 eggs
2 bags of black tea
150 grams soy sauce
1 French mini-baguette
1 piece of star anise
5 grams cumin seeds
1 orange
2 gelatin leaves (16 grams)

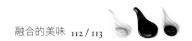

西式茶葉蛋做法
Black Tea Egg the Western Way

1. 蛋包上保鮮膜，放入鍋中，加水，以文火持續煮1小時（不要煮沸）。還有一種方法就是把蛋放到電鍋裡，加入溫水，水蓋住蛋，按下保溫功能。

2. 用一袋紅茶包泡茶，泡出濃縮的茶湯，越濃越好，所以最好事先泡。水要150毫升。

3. 柳橙取皮切丁，加入八角和小茴香籽，倒入50公克的醬油，還有50公克事先泡好的紅茶，以文火煮20分鐘。濾掉香料，留下湯汁，擱在爐上保溫。

4. 將吉利丁片泡在冷水中，一融化就倒入溫熱的醬汁中，放入冰箱冷凍1小時。

5. 打開第二袋紅茶包，把茶葉倒出來，放進果汁機裡，再倒入50公克的醬油，加入麵包，打成一團溼麵糰。把麵糰攤到烘焙紙上，用另外一張烘焙紙蓋住，送進預熱至攝氏150度的烤箱中烤到酥脆（約15分鐘）。出爐後，放到乾燥處放涼。

6. 神奇的特色上場了！紅茶、茶餅、蛋和吉利丁都備妥了，該打醬油泡了。用攪拌器打醬油，反覆地上下移動，儘量把空氣打進去，越多越好。

7. 把溫熱帶殼的蛋和液態蛋白舀到碗中。建議把蛋舀到不同的碗裡，剝好蛋再用湯匙舀回上菜的碗裡。

8. 用湯匙舀一片吉利丁，再舀一匙醬油泡。用一只小杯裝滿紅茶，茶越熱越好。最後，擺上紅茶餅裝飾，立即端出，請客人自行組合，配出美味。這就是用非傳統的手法呈現的傳統味道。

Put the eggs in a pan with water and cover them with plastic wrap. Maintain a consistent, low temperature for one hour (do not bring the water to a boil). An alternative method is to put the eggs inside a rice steamer, covered with warm water, using the keep hot function.

With 1 of the black tea bags, prepare a concentrated tea. To get the strongest taste, prepare in advance. Use 150 millilitres of water.

Dice the orange skin and combine with the star anise and cumin. Add 50 grams of soy and 50 grams of the black tea prepared in advance. Cook for 20 minutes on low heat. Sieve the spices and retain the sauce, keeping it warm on the burner.
Put the gelatin leaves in cold water and, once dissolved, add to the warm sauce. Put inside the refrigerator for one hour.

Open the second black tea bag and remove the dry leaves. Place them in a blender with 50 grams of soy sauce and the bread. Blend everything until it forms a wet dough. Spread this dough out on a sheet of oven paper and cover with one more piece of oven paper. Now bake at 150ºC until crispy (around 15 minutes). After removing from the oven, store in a dry place.

Now for the special touch! The black tea, tea cookie, egg and gelatin are all ready. It's time to prepare the soy bubbles. Using a hand blender, simply mix the sauce, trying to introduce as much air as possible by moving the mixer repeatedly in an up and down motion.

In the centre of a soup dish, serve the warm, shelled egg and some liquid egg white. I suggest that you crack the egg in a separate bowl, transferring it to the serving bowl using a spoon.

Put a square of gelatin on a spoon; fill a second spoon with some of the soy bubbles. Fill a little cup with the black tea, serving it as hot as possible. Finally, decorate the egg with the black tea cookie and serve immediately. Invite your guests to combine the ingredients themselves, reassembling these wonderful flavours. Now you have the traditional taste, but a very unconventional presentation!

鮪魚刺身＋山葵醬油凍＋味酥泡沫白蘿蔔卷

Tuna Sashimi with Soy-Wasabi Sponge, Daikon Radish Cannelloni
and Mirin Foam

　　我在台灣開餐廳之後，幾乎每天上市場買漁貨，偶爾發現上好又新鮮的鮪魚，就想為客人設計一道別處吃不到的鮪魚刺身，基本上生鮮魚肉都一樣，主要的變化在於搭配生魚片的醬油、山葵、蘿蔔絲和味酥，我把刨絲的白蘿蔔，削成薄片捲成容器，放進甜甜的味酥泡沫。

　　醬油則在加入明膠片之後用打蛋器打成有彈性的慕斯，像一塊醬油海綿蛋糕陪在生魚片的旁邊，山葵變成粉末，隨興撒在盤邊，於是所有基本組成素材都一樣的鮪魚刺身，因為我的創意發想，變成很有趣的 Fusion 美食，連日本客人吃了都非常喜歡呢！

　　I almost go to the market and buy fish everyday ever since I opened a restaurant in Taiwan. Sometimes when I find fresh tuna of top quality, it inspires me to create for the guests a special tuna sashimi dish that they won't find elsewhere. Basically, fresh fish meat is no different and the main changes are in soy sauce, wasabi, mirin and shredded daikon. I cut daikon into thin slices and roll sliced daikon up to hold the whipped sweet mirin bubbles.

　　Then, I soak gelatin slices in the soy sauce, whip the mixture into mousse, and make this soy sauce sponge cake as a side dish. As for wasabi, I ground it into powder and sprinkle on the side of the plate. As you can see, all the ingredients are the same, but with my creative imagination they become a fusion dish that even Japanese guests enjoy it so much.

Cooking time ›› 40 minutes.

主材料：

刺身等級的鮪魚	200公克	200 grams sashimi quality tuna
醬油	100公克	100 grams soy sauce
水	20公克	20 grams water
明膠	16公克	16 grams gelatin
山葵（芥末）粉	10公克	10 grams wasabi powder
味酥	100公克	1 daikon radish
大豆卵磷脂	10公克	100 grams mirin
		10 grams soy lecithin

做法：

1. 將明膠片置於容器中，放入冷水和冰塊使其軟化。

2. 在一只鍋中將醬油和礦泉水混合之後加溫，加入明膠片，並避免加入更多的水。明膠融化後，將混合好的汁液放入冰箱冷藏3小時。

3. 明膠凝固後，用打蛋器將其打至黃色慕斯狀，再放進冰箱冷藏1小時備用。

4. 白蘿蔔削成薄片，用白蘿蔔片捲成捲狀。

5. 用一把鋒利的刀將鮪魚切成方塊後，放在一個冰盤子上。將山葵醬油凍切成同樣大小，放在鮪魚的上面。撒上山葵（芥末）粉，白蘿蔔捲放在旁邊做裝飾。

6. 在碗中，用打蛋器把味醂和大豆卵磷脂打到起泡，並用最快的速度將泡倒入白蘿蔔捲，立刻上菜！

Put the gelatin leaves in a pot with cold water and some ice cubes to soften.

Combine the soy sauce and the mineral water in a pan. Heat the mixture until it´s warm, and then add the gelatin, trying to avoid adding more water. After the gelatine has dissolved, place the mixture inside the refrigerator for three hours.

After the gelatin has set, using an electric mixer, whip until the colour becomes almost yellow and the texture seems to be a mousse. Put back inside the refrigerator and wait one more hour before using.

Shave the daikon radish into thin slices. Make a roll with these slices to obtain the cannelloni.

Cut the tuna into square pieces with a sharpened knife and serve it on a cold plate. Cut the soy sponge into the same size square pieces and place them over the tuna. Sprinkle the top with wasabi powder. Place the radish cannelloni beside the tuna.

In a bowl, mix the mirin and the lecithin, whipping it until it bubbles. As fast as possible, put this bubbly foam inside the radish cannelloni. Serve immediately.

Tofumole

酪梨豆腐醬佐餛飩脆片

「Tofumole」這道食譜的原始構想來自於墨西哥的酪梨醬，我只是把中國人的豆腐加進去和蔬菜、酪梨、辛香料一起打成泥，醬裡不但多了豆腐香，入口也更稠滑柔密。

這是我為愛吃豆腐的老婆大人設計的食譜，墨西哥人用玉米片沾著酪梨醬吃，為了搭配豆腐酪梨醬的口感，我改用酥炸餛飩皮取代玉米脆片，是很棒的下酒點心和餐前開胃小點。

The idea for recipe of Tofumole originates from the avocado sauce of Mexico. I just want to add the Chinese tofu in and blend with vegetable, avocado and spice. The sauce not only has the flavour of tofu, but also thick and smooth in the mouth.

This is the recipe specially designed for my lovely wife. Mexicans like to eat tortilla chips dipped with avocado sauce. To accompany the texture of tofu avocado sauce, instead of tortilla chips, I use deep fried wonton wrappers. It's a great snack to go with wine and appetizer.

Cooking time›› 20 minutes.

主材料：

材料	數量	Ingredient
酪梨	1個	1 avocado
去皮番茄	1顆	1 peeled tomato
去皮紅蔥(法式洋蔥)	1顆	1 peeled shallot (French onion)
小茴香籽	2公克	2 grams cumin seed
鹽	5公克	5 grams salt
乾辣椒	5公克	5 grams dry chilly
蒜頭	1瓣	1 clove of garlic
嫩豆腐	50公克	50 grams soft tofu
新鮮餛飩皮	8片	8 slices of fresh won ton paste
黑胡椒	10公克	10 grams black pepper
糖	10公克	10 grams sugar
新鮮胡荽	10公克	10 grams fresh coriander

做法：

1. 將去皮酪梨、洋蔥、番茄、小茴香、大蒜、辣椒、豆腐、黑胡椒、糖及鹽全部放入食物料理機，使用手動式將所有食材打成粗泥狀，但仍看得到食物塊狀。
2. 胡荽切碎。油炸餛飩皮直到酥脆。
3. 在一個圓盤中，放上酪梨豆腐醬，撒上切碎的胡荽末，並以炸餛飩皮做為裝飾。

Put inside the blender the peeled avocado, the onion, the tomato, cumin, garlic, chilly, tofu, black pepper sugar and salt and blend it using the turbo function several times to obtain a paste but keeping some whole pieces of the ingredients.
Chop the fresh coriander as small as possible.
Deep fry the won-ton paste until is crispy.
On the bottom of a round plate serve the avocado and tofu paste, garnish with some chopped coriander and use the fried won-ton as decoration.

不一樣的牛肉麵

　　這是我的副主廚張榮恩（John）的創意食譜，他把台灣味的牛肉麵徹頭徹尾改造成另一道菜的模樣，卻在味覺中留下所有對於牛肉麵的記憶，例如充滿八角和五香味的牛肉香氣，但在造型上，炙烤過的牛肉捲上煮軟的胡蘿蔔，變成一枚壽司立正在盤中，淋上滾燙的牛肉高湯，很東方的牛肉麵，就這麼變得不一樣了！

This is a creative recipe from my sous chef John, who created another look of Taiwanese beef noodles into another course, with all the memory with regard to beef noodles left to the sense of taste, such as full of beef smell with aniseed, five spices, however, the roasted beef roll wrapped with softened carrot has become a sushi stands in the plate and sprinkled with boiling beef soup stock, an extremely eastern beef noodles has then become different.

Cooking time›› 3 hours.

主材料：

胡蘿蔔、番茄、洋蔥、韭菜	各1支
薑	2公克
大蒜	1瓣
豆瓣醬	20公克
醬油	40毫升
冰糖	5公克
白糖	10公克
牛腰脊肉	100公克
牛膝骨	1塊
水	1公升
八角、檸檬皮、肉桂、丁香、甘草粉	2公克
麵條	100公克
義大利麵條	1包

1 carrot, tomato, onion, leek
2 grams ginger
1 clove of garlic
20 grams Thick Broad-Bean Sauce
40 millilitres soy sauce
5 grams crystal sugar
10 grams white sugar
100 grams beef strip loin
1 beef knee bone
1 litres of water
2 grams star anise, lemon skin, cinnamon, clove, liquorice powder.
100 grams noodles
1 packet of spaghetti

做法：

1. 除了牛腰脊肉、麵條和半條胡蘿蔔外，其他的材料全部煮成湯，慢慢熬，熬3小時。
2. 片牛腰脊肉備用，片得越薄越好。
3. 半條胡蘿蔔切片，煮2分鐘。
4. 濾出高湯，用熬好的高湯下麵。麵煮熟撈起，用油煎到酥脆。
5. 接著準備牛肉捲：用一片煮軟的胡蘿蔔裹住一束麵條，再用牛肉片裹住胡蘿蔔，封捲，用義大利麵條當叉子串。
6. 用電鍋煮點飯，趁熱配著吃。湯要燒得熱熱的，越熱越好，淋在牛肉串上。

Prepare a soup with all the ingredients (except the beef strip loin, noodles, and half a carrot). Cook slowly for 3 hours.
Slice the beef as thin as possible and set aside.
Slice half a carrot and boil for two minutes.
Strain the sauce generated from cooking the soup stock. Cook the noodles inside this stock until ready. Remove the noodles and deep fry in oil until crispy.
Next, prepare a beef roll. Begin with a bundle of noodles. Wrap them inside a soft, boiled carrot slice. Next, wrap the beef around the outside of the carrot. Close the roll and secure with a spaghetti strand as a skewer.
Heat a little bit of rice in a steamer and serve warm. Serve the soup as hot as possible over the beef roll.

有史以來最棒的食譜
The Best Recipe Ever

　　這是我這輩子做過最棒的一道菜，這道菜的做法非常簡單，同時也是到目前為止我所做過最複雜的一道菜。

　　煮出來的東西所帶來的體驗，前所未有的刺激。這道菜的味道講起來很容易，嚐起來是甜的，口感呢，軟軟的，聞起來呢，是生命的氣味，聽起來呢，就是笑聲，看起來，就是美。

　　我替這道菜取名瑪莉娜，她是我的女兒。不過為了公平起見，我必須承認因為有大廚莫妮卡的幫忙，我才能做出這道神奇的菜。我們的功勞各半，各出了一半的力。

　　感謝莫妮卡，感謝瑪莉娜，她們是我生命中最重要的一部分，讓我心中充滿了靈感。

　　The best recipe I ever created is so simple, but -by far-it is the most complex.

　　The final product will provide you with the most exciting experience you will ever have. The taste is so easy to describe: sweet; the texture: tender; the smell: life; the sound: laughter; and the look: beauty.

　　I called this recipe MARINA—my daughter. However, to be fair, I must say that Chef MONICA helped me to create this fantastic recipe. We share the credit, each of us contributing our 50%. Thanks Monica, thanks Marina, for being the most important part of my life and for giving me the inspiration that fuels my heart.

Cooking time›› 6480 hours.

主材料：
愛心 1公克
熱情 1公克
希望 1公克
耐心 100公斤

1 gram Love
1 gram Passion
1 gram Hope
100 kilos Patience

做法：
只要把所有的材料混在一起，放入烤箱，慢慢烤，烤上6480個小時，就大公告成了！

Just mix all the ingredients above, put them in the oven, and cook slowly for 6480 hours.......et voilà!!!

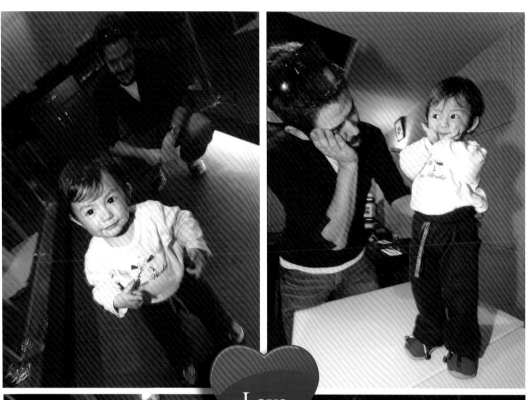

Love

主廚的一天...
A day in the life of a chef...

酒單推荐
Recommended for wine menu

蒜苗湯佐煙燻蛤蜊凍
Leeks Soup with Smoked Clams Soft Gelatin

酒單推荐：NV Jané Ventura Cava Brut

品酒筆記：被西班牙知名葡萄酒指南評鑑為最好的Cava之一，風味優雅細緻。帶有蘋果、葡萄柚的新鮮果香及花香和杏仁、草本香料，口感均衡，是極佳的開胃酒。

Tasting notes:Ranked as one of the best Cava by noted wine guide in Spain, with graceful and fine flavor, bright light yellow hue, fine bubble, with scent of fresh grapefruit, flower and almond, herbal flavor, balanced mouthfeel.

海鮮湯
Seafood Soup

酒單推荐：Cuvee Alexandre Chardonnay Casa Lapostolle

品酒筆記：一種非常濃醇的Chardonnay白酒，展露複雜及強烈的氣味；包含特殊的熱帶水果及烤布丁果香。入口有成熟水果及香料在味蕾中發散。

Tasting notes:A very rich Chardonnay, it exhibits complex and intense flavours that include tropical fruit and crème brulee aromas in a unique expression. Showing ripe fruit and hints of spice on the platate.

西班牙焦糖肉桂海綿蛋糕
Torrija

酒單推荐：Santamaria Crema Sherry wine

品酒筆記：這個製作良好的奶油色雪莉酒有著紅褐色澤，飽實甜美的果實果仁香味及鮮甜水果味，光滑柔綿的口感滿足地在口中擴散。

Tasting notes:This well-made cream Sherry has a dark mahogany color, generous sweet fruit and nut aromas and sweet dark fruit flavors with a long satisfying, smooth creamy finish.

什蔬燴蛋
Pisto (Sauted Vegetables with Egg)

酒單推荐：Non Vintage Brut Green Point

品酒筆記：混合Piont Noir 及Chardonnay兩種白酒而成，晶瑩的金黃液體帶出圓潤的花香及酵母味，蜂蜜與成熟的桃子味融合成微妙、滑順及乾淨的餘味。

Tasting notes:Made from a Combination of Piont Noir and Chardonnay, the shimmering golden yellow liquid bring out well rounded floral and yeast notes. Honey flavors on the palate, together with notes of ripe peaches, make for a subtle, smooth and clean finish.

西班牙墨汁烏賊燉飯
Squid's Ink Rice Cooked on Paella

酒單推荐：Sauvignon Blanc Semillon

品酒筆記：成熟的核果及花香味撲鼻，帶有少許橡木桶氣味，是完美的餐前酒，適合搭配海鮮及冷盤食物飲用。

Tasting notes:Ripe stone fruits and floral nose , slightly oaked；it is a perfect aperitif wine, and matches seafood and cold cut dishes well.

牛肝菌黑松露炒蛋
Scrambled Egg with Boletus Edulis and Black Truffle D.I.Y

酒單推荐：Valdrinal Sqr2

品酒筆記：採用西班牙斗羅河岸產區內40年Tempranillo葡萄老藤所摘下的少量高品質葡萄釀製而成。全球限量3000瓶，可謂為松露等級般的上等紅酒(台灣地區限量600瓶)。

Tasting notes:Made from rare high-quality grape harvested from 40-year-old vine of Tempranillo along the Toro River in Spain. With limited production of 3000 bottles worldwide, and can be proclaimed as similar-to-truffle first-class red wine (a limited quota of 600 bottles in Taiwan)

多諾斯蒂亞式蟹肉
Crab Meat "Donostiarra" Style

酒單推荐：Chardonnay Reserva Terrazas De Los Andes

品酒筆記：聞起來有熱帶鳳梨、椰子及柑橘味在鼻尖迴盪，焦糖味襯托出新鮮度，品啜之後，香草乳香的餘味久久不散。

Tasting notes:It has tropical fruits notes of pineapple, coconut and hints of citrus fruits on the nose. Subtle caramel flavours offset the freshness, leaving a creamy vanilla aftertaste on a long, lingering finish.

加利西亞章魚盤
Octopus a Feira Style

酒單推荐：Valdrinal Santamaria Verdejo

品酒筆記：採用西班牙斗羅河岸產區內生產的Verdejo青葡萄品種釀製而成。酒體細緻，酸度適中，充滿清新田園濃郁花香。

Tasting notes:A wine produced using 100% Verdejo variety grapes, which is renowned for its fresh and fruity quality. Unctuous on the palate with a pleasant first impression and a notable acidity very well integrated. Ideal accompaniment for all kinds of pasta, rice, vegetables and fish.

西班牙海鮮鍋巴飯
Spanish "Paella"

酒單推荐：Pico Madama

品酒筆記：西班牙東南部Jumilla產區海拔360公尺到720公尺左右的葡萄園種出的葡萄，口感飽滿，單寧飽滿但不澀。

Tasting notes:Created from Jumilla, southeast of Spain, with grape grown in vineyard of 360-720 meters high in altitude. Rich taste, solid tannin without astringent taste.

有史以來最棒的食譜
The Best Recipe Ever

酒單推荐：Clos Apalta 2005

品酒筆記：以老藤Merlot 及 Carmenere為主體，散發出令人愉悅的Cabernet紅酒氣味，是一款有著豐富複雜度的世界級好酒，有濃郁黑色果類並摻有少量香料及煙燻果香味，口感滑順平衡，餘味悠揚優雅，極為動人。

Tasting notes:Dominated by old vine Merlot and Carmenere, it also has a welcome splash of Cabernet. A world-class wine of great complexity, it has aromas of concentrated blackfruit with a hint of spice and smokiness. Smooth, well-balanced with a long elegant finish, this is a tremendously appealing wine.

燴煮紅鰱鮮蝦
Red Mullet and Shrimps Hot Pot "Caldereta"

酒單推荐：Non Vintage Brut Rose Green Point

品酒筆記：酒色讓人聯想到粉紅桃子外皮，有著明亮玫瑰金到淡古銅色色澤，新鮮成熟的黑櫻桃、柑橘及核果氣味撲鼻。

Tasting notes: The colour is reminiscent of pink peach skin with vibrant rose gold to light bronze hues. Fresh ripe black cherries, citrus and stone fruit notes dominate the nose.

松子雞肉卷佐摩典納甜醋
Caramelizee Stuffed Chicken with Pinenuts and Sweet Modena

酒單推荐：Cape Mentelle Chardonnay

品酒筆記：濃稠的Chardonnay白酒，帶有熱帶水果甜瓜、鳳梨及柑橘氣味，且有複雜悠長的餘味，適配海鮮類、肉類及燉煮類菜餚。

Tasting notes: A full-bodied Chardonnay with tropical fruit notes of melon, pineapple and citrus fruits. The wine is round on the palate, complex and boasts a long finish; best enjoyed with seafood, poultry and stews.

有媽媽味道的燉肉
Stewed Meat, "Mama's Style"

酒單推荐：Cabernet Merlot Cape Mentelle

品酒筆記：典型混合型紅酒，輕柔且平衡，擁有豐富黑莓香及暖暖的香草味。

Tasting notes:A classic blended red wine, it is soft and well-balanced, boasting a bouquet of lush blackberries and toasty vanilla.

肉桂牛奶燉飯
Rice with Milk and Cinnamon

酒單推荐：Michelsberg Piesporter Spatlese

品酒筆記：這是一款口感柔順且帶有豐富果香及和諧酸度的葡萄酒，可以平衡過甜的口感。

Tasting notes:Johannes Egberts Piesporter Michelsberg Riesling Spätlese is soft in taste with plenty of fruit and a harmonious tartness justly balances over sweet disserts.

西班牙蛋餅
Spanish Tortilla

酒單推荐：Red Label Claret Newton

品酒筆記：Claret是一種以Merlot酒為主體，混合Cabernet Sauvignon，Cabernet Franc，及Petit Verdot，帶有Bordeaux風格的酒類，透過濃郁的黑莓果香味展現其高雅結構。

Tasting notes: A Bordeaux style Claret is dominated by Merlot blended with Cabernet Sauvignon, Cabernet Franc and Petit Verdot. The elegant structure of the wine reveals itself through the lush blackberry aromas.

"改版"聖人骨
Saint's Bones, "Updated"

酒單推荐：Etim Verema Tardana

品酒筆記：100% 晚熟的Garnacha葡萄品種釀製，酒色明亮，美麗的紅寶石色澤中，有水果、無花果、果醬、葡萄乾和黑巧克力香，清新細緻的甜味中帶有怡人的丹寧酸，正好平衡過於甜膩的點心。

Tasting notes: Made from 100% late-maturing grape species - Garnacha, bright hue and beautiful ruby color, with scent of fruits, fig, jam, raisin, and black chocalate, a fresh and fine sweet flavor with pleasant tannin justly balances over sweet disserts.

椰子玫瑰西班牙櫻桃冷湯
Cherry Gazpacho with Coconut and Rose Blossoms

酒單推荐：The Puzzle Newton

品酒筆記：混合來自各地不同種類釀酒廠的酒類，一種技巧豐富多層次味覺的酒類，輕柔、醇厚及絲般口感，約有10~20年酒齡，是一款完美平衡的葡萄酒。

Tasting notes: So named for a blend utilizing the best plots from diverse section of the winery, the distinctive characters from each plot remains separate until it is assembled together like a puzzle by the winemaker. A finesse wine with layers of complexity; soft, full and silky, it's a well balanced wine that will happily age 10- 20 years.

烤鳳梨與辛香巧克力配蜂蜜餅乾
Toasted Pineapple with Spicy Chocolate Cookie and Rum Flambe

酒單推荐：Ten Nectar d'or Glenmorangie

品酒筆記：在波本原酒桶中熟化且在Sauternes酒大酒桶中窖藏陳釀，替這種純麥威士忌增添滑順的稠度及奢華口感。

Tasting notes:Matured in ex bourbon casks and extra-matured in Sauternes wine barriques , give this single malt whisky its smooth body and sumptuous taste.

西式茶葉蛋
Black Tea Egg "The Western Way"

酒單推荐：Malbec Terrazas De Los Andes

品酒筆記：清新的黑莓果香與花香、土壤及甘草的味道融合，創造出這款滑順及圓潤的酒款。

Tasting notes: Fresh fruity aromas of blackberries are combined with floral, earthy flavours and licorice notes to create a smooth and round wine.

鮪魚刺身 ＋ 山葵醬油凍 ＋ 味醂泡沫白蘿蔔卷
Tuna Sashimi with Soy-Wasabi Sponge, Daikon Radish Cannelloni and Mirin Foam

酒單推荐：Te Koko Cludy Bay

品酒筆記：一種豐富且風味醇厚的Sauvignon Blanc，在橡木桶裏成熟，帶有綿滑圓潤的口感，滑順的酸度餘味在口腔迴盪盤旋。

Tasting notes:A lush and full flavoured Sauvignon Blanc, matured in oak; it is creamy and ripe on palate, smooth acidity with a long finish.

椰奶蛤蜊濃湯
In Equilibrium

酒單推荐：Red Label Chardonnay Newton

品酒筆記：有著杏桃與桃子混合的誘人果香，並帶有烤麵包的香氣。入口均勻散佈
在舌尖，並遍及每個味蕾，帶有滑順怡人的酒酸。

Tasting notes:An attractive floral expression which displays a nice mix of crisp apricot
and peach aromas, with toasty notes. Opening up beautifully on the palate,
and balanced overall, it's smooth with good acidity.

熱蛤蜊紫薯沙拉佐薄荷無花果醋
Warm Clams and Purple Potato Salad with Mint
and Figs Vinagrette

酒單推荐：Unfiltered Chardonnay Newton

品酒筆記：選自葡萄園中最好的葡萄釀造，是未經過濾的酒，因此能產出細緻果香
味的Chardonnay白酒，有著香橙花、椰子、奶油糖果及烤杏仁的多層次
口感。

Tasting notes:Created from the best grapes grown in the vineyard, the wine is unfiltered,
thus yielding a Chardonnay with delicate fruit aromas. Notes of orange
blossom, coconut, butterscotch and toaster almonds unfold with layers of
complexity on the palate.

百香美乃滋藍鸚嘴魚佐魚鱗酥片
Blue Parrotfish with Eatable Scales
and Passion Fruit Mayonnaise

酒單推荐：Sauvignon Blanc Cloudy Bay

品酒筆記：一款全新的世界經典酒款，淡稻草黃的色澤，有果香、口感清爽，並帶
有濃厚的熱帶水果芭樂、杏桃、酸橙的味道，與酒酸完美調和，餘味久
久不散。

Tasting notes: A new world classic. Pale straw yellow in colour, it's aromatic, crisp and
fresh, and carries lush flavours of passion fruit, guava, apricot and lime on
palate; all of which is balanced with good acidity and a long finish.

嫩煎鵝肝&藍色檸檬蛋糕佐黑色蜂蜜
Sealed Goose Liver with Blue Lemon Cake
and Black Honey

酒單推荐：Iberia Sherry

品酒筆記：約10 -12℃冷藏飲用。

Tasting notes:Serve well chilled, about 10~12℃

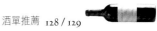

加里亞式炙干貝佐番茄粉
Sealed Scallop with Tomato Powder,
"Gallega" Garnish and Mandarin Oil

酒單推荐：Chardonnay Cloudy Bay

品酒筆記：有葡萄柚、檸檬花、腰果的味道，高雅平衡的味覺，同時帶出礦物特色
的成熟果實，餘味纖長平滑。

Tasting notes:Grapefruit, lemon flower and cashew nuts on the nose, elegant and
balanced on palate. It also offers ripe fruit with a touch of minerality, and
carries on with a long, smooth finish.

嫩鵝肝佐希臘優格及橘子綠茶魚子醬
Pate De Foie-Gras with Greek Yogurt, Orange
and Green Tea caviar

酒單推荐：Emilio Lustau Solera Sherry

品酒筆記：約 12℃冷藏飲用。

Tasting notes:Serve well chilled, about 12℃

紅胡椒煎草莓配蜂蜜餅乾
Sauted Strawberry with Pink Pepper
and Honey Cookie

酒單推荐：Late Harvest Riesling Cloudy Bay

品酒筆記：一種經特殊釀造且產量稀少的甜酒，金黃色澤有著濃郁柑橘及薑花香，
口感滑順，加上恰如其分的酸度，適合餐後享用或搭配微甜食物飲用。

Tasting notes:A sweet wine only produced in the most exceptional of vintages, and only
in very small numbers. Golden yellow colour with a rich citrus and ginger
flower nose, smooth on palate, unctuous, very good acidity; it should be
enjoyed after the meal or as a match with slightly sweet dishes.

酪梨豆腐醬佐餛飩脆片
Tofumole

酒單推荐：Amontillado Contrabandista Sherry Wine

品酒筆記：這也許是所有雪莉酒中最複雜的一款，Amontillado帶琥珀色，榛果及少
許的發酵味，微甜中帶有清新滑順的味道，且有持續而複雜的餘味。

Tasting notes:Perhaps the most complex of all Sherries the Amontillado has an amber
color, hazelnuts and hints of yeasty nose, a slightly sweet flavor with fresh
and smooth taste, lasting and complex finish.

不一樣的牛肉麵
Beef Noodles

酒單推荐：CHEVAL DES ANDES

品酒筆記：標榜混合Cabernet Sauvignon及Malbec兩種葡萄品種，有豐富技巧及複雜
度的酒款，有絕佳的熟化潛力。濃郁且高雅的味道撲鼻而來，黑醋莓、
煙草及咖啡味鮮明，帶點薄荷味。

Tasting notes: A wine of great finesse and complexity, it has excellent ageing potential.
Rich and elegant on the nose, notes of blackcurrant, tobacco and coffee are
evident along with hints of mint.

大廚愛用的好食材
Good ingredients favored by the chef

葡萄牙天然海鹽

來自葡萄牙福爾摩沙河國家公園鹽沼的天然海鹽，過度工業化使這種極佳的鹽與鹽沼幾乎絕跡，幸虧國家公園的創建使鹽沼得以留存。傳統收鹽法將海水的礦物質保留在鹽中，風味與一般工業鹽截然不同。純淨的鹽晶是由太陽曬出的天然結晶，調味時只需要輕撒一點點。

搭配食材：水煮蛋、生菜、生醃肉或魚片、薄餅芝士、鵝肝

Natural sea salt from Portugal

Sea salt from Portugal comes from the salt marshes of Ria Formosa National park, in Algarve region, in the southern Portugal. Excessive industrialization almost has made disappear this very famous salt and these salt marshes, but thanks to the creation of the natural park, the salt marshes have been preserved. Traditional gather methods allow keeping all sea water minerals in this salt, contrary to a called industrial salt. These fine salt crystals are naturally formed on crystallizing dishes thanks to the sun complicity. Sprinkle just a few for seasoning all your dishes.

Food pairings: boiled egg, raw vegetable, meat or fish carpaccio, mozzarella cheese, foie gras.

安東尼卡農頂級冷壓橄欖油

高冒煙點的特性，無論煎、焙烤、調醃料，或炸都非常適合，是種多用途料理的油品，也是令人驚豔的高價值油品！

產　　地：科爾多瓦地區、安達盧西亞、西班牙
製 造 商：安東尼卡農(Antonio Cano)。
橄欖品種：阿爾貝戈娜(Arbequina)、歐西布蘭卡(Hojiblanca)、帕加瑞納Pajarera）

Olive oil 0´4° ANTONIO CANO

An all-purpose oil that is perfect for sautéeing, roasting, for making marinades, and even for frying due to its high smoke point. A great value for incredible oil!

Region: Cordoba region, Andalusia, Spain
Producer: Antonio Cano.
Olives: Arbequina, Hojiblanca, Pajarera

無花果香巴薩米克醋

盛裝在玻璃瓶中，帶著櫻桃和無花果的天然香料與果香的巴薩米克酒醋，只需少少幾滴，就能為肉、魚、沙拉和甜點增添風味。無花果的香氣與摩典納甜酒醋和諧相融，將地中海風情完整封存在瓶裡。

搭配食材：菊苣沙拉拌核桃、帕瑪火腿佐羊奶起士串、鵝肝、嫩鴨、兔肉、洋蔥酸醬、茄片、帕瑪森起士、藍起士佐杏仁、洋梨、香草霜淇淋。

Balsanic Condiment of Modena Fig Flavored

Natural or fruity flavored with cherries and figs, just a few drops of these delicate vinegars, presented in pretty glass bottles, are enough to flavor meats, fish, salads, and desserts.

When the flavor of figs combines so harmoniously with the balsamic vinegar of Modena, it is the scent of Mediterranean that is captured in the bottle.

Food pairings: chicory salad with walnuts, Parma ham & goat's milk cheese brochette, foie gras, duckling, rabbit meat, onion chutney, eggplant slice, parmesan cheese, brousse cheese with almonds, pear, vanilla ice cream.

義大利番茄粉

義大利北方波河平原採收的熟美番茄製成的紅色番茄粉，完全天然，不含添加物和防腐劑，忠實呈現義大利豔陽下的番茄熟成香濃美味。可輕撒在菜餚上作為最後的調味，也可以添加水分和橄欖油做成醬料。或代替新鮮番茄做為菜餚裝飾。

搭配食材：沙拉、吐司、湯品、水煮蔬菜、義大利麵、調味醬、裝飾等。

Italian tomato powder

This powder made from ripe tomatoes harvested in summer in the fertile Pò valley of the northern Italy allows to find the intensity and freshness of ripe tomatoes all over the year. Totally natural, it contains no additives and no preservatives. All the concentrated flavor of tomatoes ripped in the Italian sun.

This tomato powder can be sprinkled like a spice to add a final touch to all your dishes. It is also an excellent sauce basis when hydrated with olive oil. It can also be used as a substitute for fresh tomatoes in many preparations and to garnish your plates . **Food pairings:** salads, toasts, soups, cooked vegetables, pasta, sauce basis, garnishing, etc.

特製薄荷橄欖油

一個由橄欖油與薄荷葉所混合製成的調味油，最大特色是採集自義大利普雅的新鮮薄荷，薄荷採收後幾個小時內就要與橄欖油一同壓榨。製造方式有二種：冬天，用石磨一起壓榨薄荷葉和橄欖油。在夏季，則是將薄荷葉浸泡在橄欖油中。

如何使用：淋在麥粒番茄生菜沙拉、北非小米飯、雞豆沙拉、羊肉、番茄佐羊乳酪義麵沙拉。使用在甜點上一樣相當美味：淋在梨和芒果，也可以加在布朗尼料理粉中。

Mint Specialty Olive Oil

A rare oil made from olives and mint leaves. For this specialty, the fresh mint is grown and gathered from the Galantino farm in Puglia, Italy. In the hours that follow the picking of the mint, the plant is pressed with the olives in the press. The production is in two steps. During winter, mint leaves and olives are pressed together with millstone. In summer time, other leaves infuse in olive oil.

Uses: Drizzle over tabouté, couscous, chickpea salad, lamb dishes, tomato and feta pasta salad. Also delicious for desserts: drizzle over pineapple and mango, even in brownie mix.

摩典納酒醋

O&CO 的熱銷產品，完全依循古法釀造，也是義大利摩典納酒醋中的最高品質，以傳統工序製作，巴薩米克酒醋在橡木桶中慢慢熟成，直到甜酸達到完美平衡，如蜜糖般的稠厚質地，和木桶中長時間熟成精萃出巴薩米克酒醋特有的香氣。

如何使用：淋在烤過的肉、雞肉、或魚肉、烤蔬菜，甚至甜點。

Premium Balsamic Vinegar of Modena

O&CO. Best Seller. Made according to an ancient recipe, this superior quality balsamic vinegar of Modena, Italy, has been slowly refined in a variety of precious wooden casks until it has reached a perfect balance between sweet and sour, with a syrupy texture.

Uses: Drizzle on grilled meat, chicken, or fish, baked vegetables and even desserts.

特製辣椒橄欖油

將來自智利或卡拉布里亞的新鮮的辣椒和橄欖一同壓榨，在榨取過程中添加了檸檬成功地平衡了辣椒的嗆辣，擁有辣椒的香氣而不會過度改變料理風味。只需微量淋上辣油就能提味。

搭配食材：披薩、義大利麵、烤肉、炒蛋、調味料……

Chili Pepper Specialty Olive Oil

Fresh chili peppers from Chile or Calabria and olives from the region are pressed together. The millman succeeds to balance hot notes of chili pepper adding fresh lemons during the pressing. It has the spicy flavor of the chili without overpowering your food. Spice up your dishes with a drizzle of the chili oil.

food pairings: pizza, pasta, grilled meat, scrambled eggs, sauces...

黑松露油

OLIVIERS & CO. 將來自義大利皮耶蒙的黑松露 (及白松露) 與來自義大利南方普雅的橄欖油融成一小瓶，只需幾滴便足以將冷盤或熱料理帶來更深層的美味。

如何使用：少量地淋在義大利麵、蔬菜料理、蘑菇飯、干貝、蛋和蘆筍來點綴成豪華料理。

Black Truffle Oil

This Italian specialty combines extra virgin olive oil from Puglia with black truffle aroma from Piemont (Tuber Melanosporum). Used as a final touch, this oil brings a delicious depth of flavor to your cold or hot dishes.

Uses: Drizzle over pasta, vegetable dishes, mushroom risotto, scallops, eggs and asparagus for a luxurious finish.

好料哪裡買？
Where to buy those good ingredients

利豐蔬菜行
台北市110信義區虎林街101巷1號
電話：02-2762-0419
傳真：02-2764-8814
新鮮這個詞正好用來形容這家店。這裡是蔬菜的天堂，每天提供消費者最好品質的本地食材，還有進口食材。

MORI VEGETABLE SHOP
NO. 1 Lane 101, Hu-lin Street；Xinyi
District Taipei, Taiwan, R.O.C.
Tel: 0922-270776 / (02) 2762-0419
/ (02) 2764-8814
Fresh, this word will describe this shop.
They provide us with the best quality
everyday, local ingredients and also
imported items, paradise for vegetarians....

泰奧菲Oliviers & Co.
台北市106大安區信義路三段75-1號
電話：02-2704-3077
傳真：02-2704-3055
網址：http//www.theophile.com.tw
這家店為真正的老饕提供全球獨一無二的產品和禮品。

產品：義大利摩典納葡萄酒醋、義大利番茄粉、葡萄牙天然海鹽、巴薩米克無花果醋、黑菌松露醬、0'4 ANTONIO CANO橄欖油、黑松露油、辣椒特製橄欖油、薄荷特製橄欖油、頂級巴薩米克酒醋

Glenmorangie Nectar D´OR Théophile Co Ltd
NO.75-1 ,Sec 3, Shin-Yi Road, 106 Taipei, Taiwan,
R.O.C. Tel: 886-2-2704-3077 FAX: 886-2-2704-3055
http//www.theophile.com.tw
This store offer unique products in the world and
unique gifts for real gourmets.

OLIVIERS & CO.®
LES HOMMES DE L'OLIVIE

立昇洋行
台北縣新莊市新樹路268巷18號
電話：02-2208-0669, 2203-6825
網址：http://www.cashinn.com.tw
在這裡可以找到市場上最好的西班牙酒、完美的服務、高品質和合理的價格，是該公司的理念。此外，他們也賣西班牙橄欖油。

Cash Inn Distribution LTD.
No. 18, Lane 268, Sin Su Road, Sin Tsung
City Taipei Hsien, Taiwan R.O.C. Tel:
/ (02) 2208-0669 / (02) 2203-6825
http://www.cashinn.com.tw
The company to found some of the best
Spanish wines in the market, perfect service,
good quality and good price. The philosophy
of Cash Inn. They also provide Spanish
olive oil. Valdrinal SANTAMARIA Rueda
verdejo Valdrinal SQR² Ribera del Duero
D.O.

德國福維克集團台灣分公司
台北市100中正區忠孝東路一段85號5樓
電話：02-2397-1333
傳真：02-2397-1211
網址：http://www.vorwerk.com/
福維克的美善品多功能料理機TM 31是款一機多用途的廚具，可在非常短的時間剁、切、攪、拌、碾、研、磨、揉、乳化和秤重。除了無法自行清理之外，蒸、煮、炒樣樣都行。

Vorwerk Lux (Far East) Limited, , Taiwan
Branch5F, No. 85, Sec. 1, Chung-Hsiao E. Rd.,
Taipei City 100, Taiwan R.O.C
Tel: 886-2-2397-1333 Fax: 886-2-2397-1211
http://www.vorwerk.com/
The Vorwerk Thermomix TM 31 is a multifunctional
kitchen machine that can do almost anything in no
time at all: weighing, chopping, mixing, stirring,
cutting, grating, grinding, pulverising, kneading
and emulsifying. It cooks, simmers and all but cleans
itself.

桂英 Ke Shian
台北環南市場丁棟1200號
電話：02-2309-7593 /0930483048
台北最新鮮的魚貨和海鮮，每天來自台灣最大的幾個漁港進貨。不妨試試他們的自製特產。

No.1200 Huanan Market, Taipei, Taiwan, R.O.C
Tel:02-2309-7593 / 0930483048
The most fresh fish and seafood in Taipei, coming
everyday from the biggest fishing ports in Taiwan,
try also their home made specialities.

ESTATES & WINES COLLECTION
Moët Hennessy
TAIWAN

藍象皇家泰國美饌

台北市115南港區南港路三段130巷3弄13號
電話：02-2789-2298
傳真：02-2789-0689
網址：http://www.blueelephant.com.tw
消費者可以在位於Sogo百貨復興館地下三樓的
這家店，找到最好的食材，他們總是會顧到品
質。在全世界擁有好幾家連鎖餐廳，提供一系
列的泰國菜，還可以外帶。產品包括：椰漿、
酸橙、檸檬香茅、辣椒醬。

BLUE ELEPHANT
Royal Thai Cuisine

Thai Gourmet Ltd. No.13, Aly. 3, Ln. 130, Sec.
3, Nangang Rd., Nangang District Taipei 115
Taiwan (R.O.C)
Tel: +886 2 2789 2298 Fax: +886 2 2789 0698
http://www.blueelephant.com.tw
The best Thai ingredients you'll ever found,
always taking care of quality, with several
restaurants around the world and a fantastic range
of Thai specialities. A delight for gourmets. They
also provide fresh take away food, on their store
located on SOGO B3 Fushing. Coconut Cream
Kaffir Lime Lemongrass Chili Sauce

夏朵菸酒股份有限公司

台北市大安區復興南路一段279巷8號1樓
電話：02-2708-2567
傳真：02-2708-2874
網址：www.chateaux.com.tw
台灣重要的葡萄酒進口商，進口法國葡萄酒、
頂級酒莊的葡萄酒和全球各地上好的葡萄酒，
包括美國、義大利、西班牙、德國、奧地利、
澳洲、紐西蘭、智利等國的葡萄酒。

夏朵 Chateaux

CHATEAUX WINE & CIGAR
www.chateaux.com.tw Tel: (02) 2708-2567
Fax: (02) 2708-2874
One of the most important importers for French
wine, grand crus and top global wines in Taiwan
.They also provide wines from United States,
Italy, Spain Germany, Austria, Australia, New
Zealand, Chile and many other countries.

酩悦軒尼詩－路易威登集團

（Moët Hennessy - Louis Vuitton, LVMH Group）
是一法國酒業與高價奢侈品製造集團，是貝爾
納阿爾諾（Bernard Arnault）於1987年時，將全
球著名的皮作公司路易威登（Louis Vuitton）與
酒業家族酩悦軒尼詩（Mot Hennessy）合併而
成，旗下擁有50多個品牌，是當今世界最大的
精品集團。集團主要業務包括以下五個領域：
葡萄酒及烈酒（Wines & Spirits）、時裝及皮革製
品（Fashion & Leather Goods）、香水及化妝品
（Perfumes & Cosmetics）、鐘錶及珠寶（Watches
& Jewelry）、精品零售

LVMH Moët Hennessy • Louis Vuitton S.A.
(Euronext: MC), usually shortened to LVMH, is
a French holding company and one of the world's
largest luxury goods conglomerates. It is the parent
of around 60 sub-companies that each manage a
small number of prestigious brands. These daughter
companies are, to a large extent, run autonomously.
The group was formed after mergers brought
together champagne producer Moët et Chandon
and Hennessy, a leading manufacturer of cognac. In
1987, they merged with fashion house Louis Vuitton
to form the current group.

Lacor集團

DD兄弟國際股份有限公司
台北市中山北路三段55巷30號1樓
電話：02-2586-9889
傳真：02-2586-9886
網址：http://www.lacor.es
終於有一家製造商可以滿足美食家的需求了。這
家公司提供一系列令人滿意的產品，不僅可以在
家用，就連五星級飯店也適用。商品範圍涵蓋各
種形狀與大小的廚具、廚房器皿、餐具、糕點用
具，甚至還有茶和咖啡。

LACOR MENAJE PROFESIONAL S.L.
GRUPO MARCOS LARRAÑAGA Y CIA.
DD Brothers Int'l Corp.
1F, No 30, Lane 55 Sec 3 JungShan N. Rd. Taipei
104, Taiwan R.O.C.
Tel: 0932-250561 / (02) 2586-9889
FAX: (02) 2586-9886
http://www.lacor.es
At long last there is a manufacturer who can fulfil the
desires of gastronomes and connoisseurs. Lacor offers
a range of pleasing products which can not only be
used domestically but also in a five-star hotel. As well
as covering different shapes and sizes of kitchenware
our range also includes kitchen and table utensils,
pastry ware and tea and coffee services.

西班牙大廚 到你家

A Walk Through Spain

國家圖書館出版品預行編目（CIP）資料

西班牙大廚到你家 / 丹尼爾・雷格尼亞
(Daniel Negreira)作；李瓊絲翻譯.--
初版 . -- 臺北市：橘子文化，2014.06
面 ；　公分
中英對照
ISBN 978-986-364-011-0（平裝）

1. 食譜 2. 西班牙
427.12　　　　　　　　　103010510

作　　者　丹尼爾・雷格尼亞 Daniel Negreira
英文編輯　Geoffrey Paul Carpenter,Ph.D.
翻　　譯　李瓊絲
攝　　影　陳牆
發 行 人　程安琪
總 策 畫　程顯灝
編輯顧問　錢嘉琪
編輯顧問　潘秉新

總 編 輯　呂增娣
執行主編　錢嘉琪
主　　編　李瓊絲、鍾若琦
編　　輯　吳孟蓉、程郁庭、許雅眉
編輯助理　張雅茹
美術主編　潘大智
行銷企劃　謝儀方
出 版 者　橘子文化事業有限公司

總 代 理　三友圖書有限公司
地　　址　106台北市安和路2段213號4樓
電　　話　(02) 2377-4155
傳　　真　(02) 2377-4355
E - mail　service@sanyau.com.tw
郵政劃撥　05844889 三友圖書有限公司

總 經 銷　大和書報圖書股份有限公司
地　　址　新北市新莊區五工五路2號
電　　話　(02) 8990-2588
傳　　真　(02) 2299-7900

http://www.ju-zi.com.tw

三友圖書
友直 友諒 友多聞

初　　版　2014年 06月
定　　價　新臺幣 420 元
I S B N　978-986-364-011-0 (平裝)